Neural Nets: Applications in Geography

The GeoJournal Library

Volume 29

The titles published in this series are listed at the end of this volume.

Neural Nets: Applications in Geography

edited by

BRUCE C. HEWITSON

Department of Environmental and Geographical Science, University of Cape Town, South Africa

and

ROBERT G. CRANE

Department of Geography and the Earth System Science Center, Pennsylvania State University, U.S.A.

SPRINGER-SCIENCE+BUSINESS MEDIA, B.V.

A C.I.P. Catalogue record for this book is available from the Library of Congress.

ISBN 978-0-7923-2746-2 ISBN 978-94-011-1122-5 (eBook)
DOI 10.1007/978-94-011-1122-5

Printed on acid-free paper

For

Nancy and Janet
Linda, Duncan, and Ann

Table of Contents

Preface

In many ways, the inspiration for this book can be traced back to a flamboyant seminar given by Stan Openshaw, one of the contributors to the present work. During a guest lecture series at The Pennsylvania State University, he presented the concept of neural nets to the unwary geography department--a concept that proved to be highly captivating and prompted some of us to disappear into the computer room and try our hand at a neural net program.

Not surprisingly, after the enthusiastic start, it became apparent that neural nets were not to be the panacea for all research problems. However, it was also clear that the approach offered a fascinating new strategy with enormous potential for many tasks in geographical and spatial sciences. Unfortunately, for one reason or another, neural nets have received little exposure in many research areas that could profitably use the technique. This, then, is the purpose of the book--not to provide an in-depth reference on neural nets, but to promote the concept through a working introduction, using applied examples in the human and physical geographical sciences to demonstrate the effectiveness of the technique. It is aimed particularly at exposing neural nets to researchers who have not come across the technique, or who have perhaps dismissed neural nets as uninformative black boxes. In no way is this a comprehensive survey of the field, nor does it cover all the permutations and mathematical extensions; however, we hope that it will bring the technique into wider use in the geographical (in the widest sense of the word) fields.

Chapter One

LOOKS AND USES

Bruce C. Hewitson and Robert G. Crane

Amongst many researchers the mention of neural nets evokes widely differing responses, from enthusiastic support through to emphatic dislike, or simply blank looks of non-comprehension. The reasons for this are varied, but are in part due to mis-understanding, mis-representation, and in some cases lack of exposure. For many years neural nets have also suffered from the affliction of a confusing nomenclature. For example, 'neural net' goes under such diverse names as connectionist models and neuromorphic systems. Having become rapidly popular in multiple disciplines simultaneously, and with numerous permutations, extensive sets of terms have developed that makes for confusing reading to the uninitiated. Furthermore, the literature is spread over a very broad spectrum, although the number of dedicated journals and books is rapidly growing. In this book we endeavor to provide an entry point through which researchers in the physical and social 'geographic' sciences can see the potential for their own work, and obtain a working start in the methodology. The term 'geographic' does not mean that this book is focused at that hard-to-define breed of scientists who may call themselves geographers. Rather it is meant to convey the idea that neural nets have application in the broad spectrum of the spatio-temporal relationships of the world around us.

This first chapter, then, will simply take an overview picture of neural nets both internally and externally. At the same time, it will consider conceptually how to apply neural nets, while raising some caveats and cautions. The mathematical details and application particulars are left to the later chapters.

1.0 Origins and Growth

Neural nets (NN), or, more accurately, artificial neural nets (ANN), were inspired by perceptions of how the biological counterpart--the brain--was supposed to operate. In many ways ANNs have diverged from this initial objective and have become for many an applied mathematical technique with coincident biological terminology. In practice nowdays it can be said that the ANN only represents the brain at the most elementary

B. C. Hewitson and R. G. Crane (eds.), Neural Nets: Applications in Geography, 1–9.

level of process, although the ANN has retained as primary features two characteristics of the brain: the ability to 'learn,' and to generalize from limited information. Attempts at modeling the activity of biological neural systems began as far back as the era of the Second World War, and over the years a number of papers and books appeared following this line of research. However, the practical uses for neural nets have only evolved in the last decade, which belies their long history. ANNs did not become a viable prospect until two basic problems could be overcome. The first was the dilemma of how to train the net and get the net to learn in order to solve some problem. Initial steps toward dealing with this began in the early 1950s; however, in the years that followed, ANNs, for reasons that were inexplicable at the time, seemed able to solve some problems while being unable to solve others that could be considered very simple or even trivial. Consequently, the use of ANNs fell into disfavor. The second problem was the lack of suitable computers to handle the computationally intensive task of training a net. With only a few researchers quietly working over the years it was not until the 1980s that new learning algorithms and a good theoretical framework came about, which, along with faster computers, sparked a dramatic resurgence of interest in neural nets, and many applications were found for the now viable procedure. Subsequently, neural nets have spread through the disciplines in an almost biological manner, and new algorithms and applications, or at least variants of older ones, appear with almost every new neural net paper published.

Today there are over 15 journals dealing with neural nets and as many, or more, in related fields. There are many public domain computer programs and ANN computer packages, as well as a number of commercial offerings. On Netnews, the international electronic discussion board on internet, the neural net category sees tens of discussion notes and questions posted daily. At least seven neural network related associations exist, and each year numerous conferences and workshops take place, while, more importantly, ANNs are being applied to new applications and used across the disciplines to deal with previously problematic issues. Uses of ANNs now span a huge spectrum of applications, encompassing such diverse implementations as optical character recognition, image classification, extraction of atmospheric profile information, weather forecasting, AIDS prediction, house price prediction, bomb detection, credit card application screening, cancer detection, speech recognition, playing the stock market, and even wine tasting. In the appendices we provide some reference material on where to find more on journals, organizations, and software sources.

1.1 Conceptual Overview

Before getting into the detailed mathematics and the foundations of neural nets in Chapter 2, we present a brief conceptual overview of how neural nets operate. The concept of ANNs is itself quite basic and has many analogies in the statistical world that capture one or other aspect of the ANN. However, the ANN goes further than these techniques in many areas. Consider for the moment a net as a black box with inputs and outputs, which performs some function for mapping the input to an output. Initially the net is in a random state, 'untrained,' and represents a random function. The first step then is to train the net to learn some mapping/relationship between the input and output. This is accomplished through presenting the net with a sample of known inputs and outputs, which, in conjunction with a learning algorithm, modifies the internal function performed by the net to find a relation between the input and output. Thereafter, as long as one remains within the bounds of the training samples, the net can be applied in a similar manner to further unseen data. The inputs and outputs to the net can be binary or continuous, or any mix of the two. For example, multispectral pixel information from a satellite image may form the input to a net, which could be trained with binary outputs, each of which represents a classification of land surface type; alternatively, with demographic information forming an input, the output could be trained to represent predicted AIDS populations.

Continuing for the moment with treating the ANN as a black box, for an ANN with continuous (non-binary) inputs and outputs, a useful analogy is to consider the net as a super-form of multiple regression--in fact, linear multiple regression is a special case of neural nets. In the same way as one performs regression between $\{X\}$ and $\{Y\}$, developing a relation such that $\{y\} = f\{x\}$, so does a net find some function $f\{x\}$ when trained. The attractiveness of ANNs lies in part in their ability to, in theory, represent any arbitrary non-linear function. Whereas in regression one is tied to a linear relationship, or at best a pre-specified non-linearity, the neural net finds its own function as best it can, given the complexity used in the net, and without the constraint of linearity.

Alternatively the ANN could be used in an analogous fashion to cluster analysis. From this perspective the net finds relations between different input samples that group them into classes. On presentation of an input, the net would use binary outputs to indicate which cluster the input mapped to. Continuous value outputs could even be used to give a degree of membership for each cluster. Again linearity is not a constraint, and both the analog of supervised and unsupervised 'clustering' is possible.

Other attractions of ANNs compared to their more traditional counterparts are the ability to generalize a relationship from only a small subset of data, to remain relatively robust in the presence of noisy inputs or missing input parameters, and to adapt and continue learning in the face of changing environments. In short, the ANN is highly flexible. Finally, the net is not so much a black box--an unfairly acquired reputation--as a 'grey' box, and techniques are available to interpret the function it represents.

1.2 Neural Net Structures

Again, while Chapter Two will deal in detail with the mathematics and topology of different nets, we give here a basic outline of ANN structure--the internals of the grey box--to provide an introductory overview.

Looking inside the grey box of the previous section it is easiest to use the biological analogy again. Being initially models of biological neural systems, ANNs have in common with the brain the same basic macro-structure. In the same way as the brain is composed of interlinked processing elements (neurons), so the ANN has simple processors (nodes) connected by weighted links. Figure 1.1 shows what a simple node in an ANN might look like.

A node in the net sums the weighted inputs from the links feeding into it, and performs some function on the summed value. This function is typically a non-linear bounded function, although a wide variety of options exist, including threshold functions, binary functions, and linear functions. The output is then given to further weighted links leading to other units, and many nodes may be connected together to form a net of processing nodes. Figure 1.2 shows a simple configuration, termed feed-forward, in which nodes (with possibly different functions) are connected such that the information flows unidirectionally from input to output. The net as a whole will have some input point(s) and some point(s) of output.

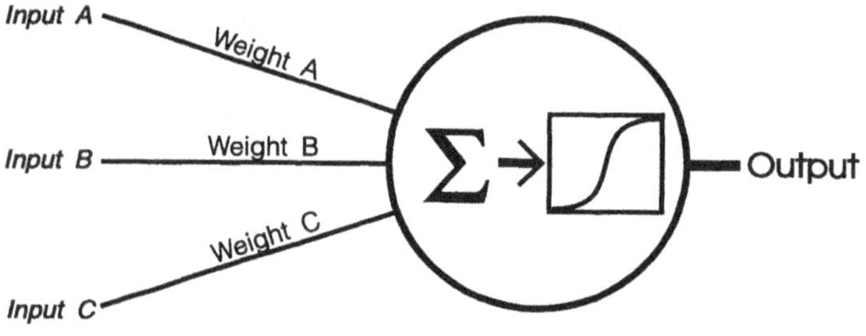

Figure 1.1: Example of a node within a Neural Net.

Typically the nodes are placed in layers, with an input layer and output layer of nodes, and usually one or more hidden layers (in the sense that they have no connections external to the net). The example shown here has one hidden layer, and has full connectivity between layers. As can be easily imagined, great complexity is available, with the possibility for links to leapfrog layers, connect to nodes within the same layer, and even feed back to earlier layers. In this way an amazing degree of complexity can be attained. Furthermore, the nodes in the net do not need to be all the same, but can be of mixed function types and contain such extra features as inhibition inputs, time-delay internal loops (memory!) and almost anything one may wish.

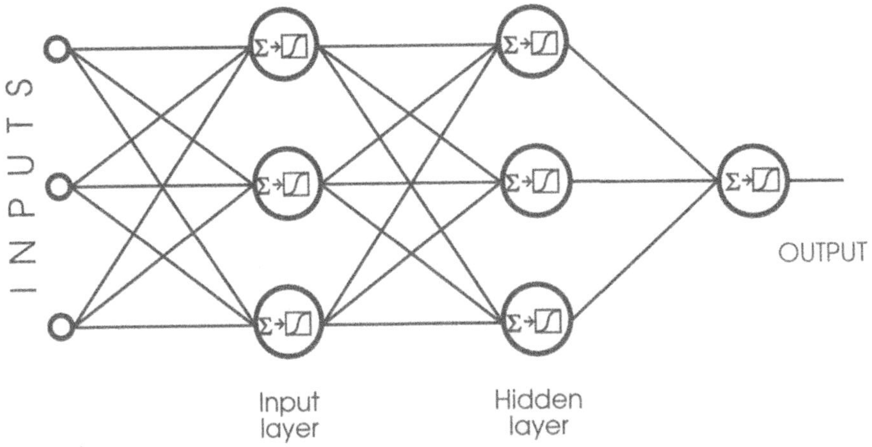

Figure 1.2: Example of a simple 'feed-forward' Neural Net.

All this versatility provides a complex manner in which to manipulate the input data. For example, in Figure 1.2 there are three inputs and one output. In trying to relate the input to the output one could try multiple regression of the form:

$$output = a1 * (input\ 1) + a2 * (input\ 2) + a3 * (input\ 3) + C \qquad 1$$

Here $a1$, $a2$, and $a3$ are the regression coefficients that weight the inputs in a fixed linear manner to produce some output. If the ANN of Figure 1.2 is employed, each input to the net is passed through 14 weights (analogous to regression coefficients), and modified by seven functions of user-specified characteristics: non-linear, threshold, binary, product, etc. Given enough nodes in the net it is theoretically possible to represent any arbitrary non-linear function. One way to think of this is to draw a (limited) parallel with inverse fourier transforms. In a similar manner to the way inverse fourier transforms combine sine and cosine waves to create a complex waveform, so the ANN combines the transfer function of the individual nodes in different ways to represent an overall complex transfer function.

1.3 Implementing the Neural Net

Constructing the neural net is only half the task. Once an internal structure is established, one still has to train the net, adjusting the weights on the links that control the flow of information in order to represent some meaningful function. Numerous algorithms exist to do this, but all come down to some form of minimization routine. Considering that the link weights within the net are what adapt the function of the net, for *n* weights one has an *n* parameter space to operate in. For any given set of weights the output of the net will fall on an error surface, and the training function is to find that set of weights in parameter space such that the minimum of the error surface is found--the global minimum.

This is the stumbling block for many of the training algorithms: in performing a minimization exercise they may become trapped in a local minimum. The simplest of all these algorithms, a gradient descent approach where the weights are continuously modified in the direction of minimizing the error, is especially susceptable to this problem. However, numerous work-around and other algorithms exist, and today one has a choice of a dozen or more commonly used methods that can be employed, with variations on the theme in almost every new paper published.

The training is commonly applied with a training set of data where the net is presented with a set of inputs and known associated outputs, and 'learns' or minimizes the error in its output for a given input and target output. As already noted, for a net of enough complexity it is theoretically possible to learn a relationship of arbitrary complexity and, in doing so, relate an input and output data set where no real relationship exists--a case of overtraining. Thus it is common to retain a portion of the data as a test data set, where, during the training process, the net is tested on independent data without training taking place. When the point is reached where no improvement is found with respect to the test data, then further training would start to learn 'noise' or features that are not really present in the data.

Application of a neural net is also possible when no target output data are available, and in this form the net resolves its own view of the structure of the input data set in a weak analogous manner to cluster analysis. In this form the net looks for ways to relate/cluster different input cases so as to organize and find structure within the input space. The structural detail that can be revealed in this case is determined by the way the net is constructed, in the same way as the complexity of a function between an input and output data set is dependent on the complexity of the net.

Once trained the net can be applied to new data (while recognizing the constraints discussed below). Moreover, while trained on some initial data set, it is also possible to use the net and continue training at the same time, continuously adapting the net to an ever changing situation.

1.4 Inevitable Caveats and Cautions

Not surprisingly neural nets are not the ultimate solution for all relational/clustering/categorization techniques, and have their own set of limitations. The most fundamental criteria is, of course, that only quantifiable data can be managed, not qualitative categorical information. This, however, still leaves a broad field of applications.

Perhaps the most important constraint to highlight is that the net can only operate on data that fit within the bounds of the training set, and that the data must form part of the same system represented by the training data. Using the net for cases that fall significantly outside the boundaries of the training data--for example, trying to forecast a system which is rapidly monotonically increasing--is liable to produce problematic results. This is not to say forecasting is out, since forecasting a system that operates within the bounds of the training is quite valid.

To operate as such, however, presumes the fact that enough training data were available to adequately span the complexity of the input space, i.e., to adequately represent all the possibilities such that the net can learn them. In the absence of this it may arise that the net is presented with an input that, while falling numerically within the bounds of the training data, nevertheless is positioned in input space at some point never learned by the net. In this case the behavior of the net is uncertain. Of course the smoother the function to be learned, the fewer the training cases that may be required, and the simpler the complexity of the net.

Another caution is that the net learns the function of the training data, which is not necessarily the function desired if new unseen data are to be used with the trained net. Even where a test data set (one not used to train the net) is used to terminate the training at some point, the net is still trained to the point that it performs the best on the test data. Where the net sees only cases used in training, such as using the net to encode or decode binary data that fall within a finite number of permutations, this is of course not a problem. However, in applications in the physical world this is unlikely to be the case, and the caution must be noted not to use the net outside the bounds of the training set.

Being trapped in a local minimum during training is also a potential problem, and care must be taken in the choice and monitoring of the training algorithm to guard against this happening. While finding a local minimum close to the global minimum may be acceptable, a local minimum far from the global minimum may not only lead to poor results, but to very misleading interpretations of the function represented. Again the smoother the function the less the likelihood of not finding the global minimum.

1.5 Where Next?

The rest of the book may be read in any order, depending on what one is looking for. Chapter Two provides the mathematical foundations that enable one to construct and train basic nets, and details some of the deeper complexities behind their operation. The bulk of the book, however, is devoted to descriptions of neural net applications in such broad-ranging fields as census analysis, predicting the spread of AIDS, describing synoptic controls on mountain snowfall, examining the relationships between atmospheric circulation and tropical rainfall, and the remote sensing of polar cloud and sea ice characteristics. The text illustrates neural nets employed in modes analogous to multiple-regression analysis, cluster analysis, and maximum-likelihood classification; not only are the neural nets shown to be equal or superior to these more conventional methods, particularly where the relationships have a strong non-linear component, but they are also shown to contain significant explanatory power. Several chapters demonstrate that the nets themselves can be decomposed to illuminate causative linkages between different events in both the physical and human environments.

Stan Openshaw's chapter (Chapter Three) is a call for a new quantitative geography based on computational methods and encompassing the full suite of Artificial Intelligence (AI) tools. Openshaw sees the field of neural computing as opening the door to this new quantitative revolution. In this chapter he discusses several different neural net architectures, but focuses attention on one particular scheme--the Kohonen Self-Organizing map (SOM)--and expands upon its utility and benefits for problems of spatial classification. Openshaw reports the results of two studies using this net: one produces a modest improvement to the grouping of census data (compared to a K-Means clustering procedure), while the other results in a substantial improvement over the more traditional classification methods. In Chapter Four, Kevin Winter and Bruce Hewitson provide an example of how the same neural net architecture has been utilized in an analysis of South African census data. Rather than seeking discrete, and discontinuous groupings of cases in a multivariate measurement space, the SOM is shown to produce a two-dimensional mapping of each variable as a continuous function, where potential groups may be defined by the intersection of variables, or combinations of variables, on this two-dimensional surface. This application demonstrates clearly the flexible nature and the 'more natural' feel of the resulting groupings.

Stan Openshaw also lists several reasons why one might turn to neural nets to analyze spatial relationships. The one that, presumably, will interest researchers the most is the possibility of obtaining greater explanatory power through the use of this technology. While we can't claim that neural nets will always give better results than other methods, we can demonstrate some very clear cases where this is so. David McGinnis shows, in Chapter Five, that neural nets provide a much superior predictive model for relating mountain snowfall in the Colorado Basin of the western United States to the large-scale atmospheric circulation. While we recognize from our understanding of physical processes that such circulation controls should exist, the clear implication from his work is that they are highly non-linear, and that the true strength of these relationships is masked by the linear filters imposed by more conventional correlation-type analyses.

A further objective of this book, beyond simply demonstrating that neural nets may give superior results for certain types of applications, is to show that the reputation neural nets have as uninformative 'black boxes' is not necessarily valid. In Chapter Six, Peter Gould uses a simple neural net to predict the 'next map' in the spread of AIDS in Ohio (USA). We use a similar net in Chapter Seven to examine relationships between the atmospheric circulation and rainfall in the tropical convective regime of southern Mexico. The significant aspect of these studies is not the ability of the nets to carry out the tasks at hand (although they are very successful in both cases), rather it is the development and application of a 'sensitivity parameter' that allows us to delve into the net and extract useful information on social and physical processes.

Finally, in Chapter Eight, Jeff Key illustrates several applications of neural nets to image analysis. Possibly the most significant outcome of this work being that neural nets can attack problems (such as the extraction and generalization of linear or curvilinear features) that are difficult using more traditional techniques, and that even where the nets do not give improved classification performance (when compared to a maximum likelihood classifier, for example) they achieve very similar results with a considerably smaller and more variable training set.

Neural nets offer a fascinating new strategy for spatial analysis, and their application holds enormous potential for the geographic sciences. However, the number of studies that have utilized these techniques is limited. This lack of interest can be attributed, in part, to lack of exposure; to the use of extensive and often confusing jargon; and to the misapprehension that, without an underlying statistical model, the explanatory power of the neural net is very low. Demonstrating a wide variety of neural net applications in geography, the following chapters attack all three issues in a simple manner and with minimal jargon.

Bruce C. Hewitson, Dept. of Environmental and Geographical Science, University of Cape Town, Private Bag, Rondebosch 7700, South Africa

Robert G. Crane, Dept. of Geography and The Earth system Science Center, The Pennsylvania State University, University Park, PA 16802, U.S.A.

Chapter Two

NEURAL NETWORKS AND THEIR APPLICATIONS

Eugene E. Clothiaux and Charles M. Bachmann

2.0 Introduction

The current interest in artificial neural networks can be attributed, in part, to the development of the modern computer. Since the advent of inexpensive, efficient, high-storage capacity computers, there has been an information explosion in many scientific disciplines as researchers are able to acquire larger and more comprehensive data sets. The interpretation of much of these data often requires manual inspection by scientists, especially when traditional methods of analysis do not appear to find the important relationships in the data. Manual inspection of data can be repetitive, time consuming, and difficult when many variables are involved simultaneously. Several novel processing schemes have been devised that attempt to supplement traditional signal processing techniques in difficult applications. One such approach for finding interesting relationships in multivariate data is the field of artificial neural networks (ANN).

The name 'artificial neural network' is derived from the similarity between the processing element used in such networks and neurons found in the cortex of animals. The basic processing element in an ANN, variously called a node, unit (our preference), neuron, or adaline (adaptive linear element), has six essential parts (Table 2.1, Figure 2.1): n fibers that convey the inputs to the unit, n weights that receive the n inputs, an optional bias weight with a constant input of one, a function u that combines the inputs and weights to produce the unit's activation, a function σ that operates on the unit's activation to produce the unit's output, and a branching output fiber that relays the unit's output to other units in the network.

Table 2.1: Elements of an artificial neural network unit.

Elements	Function
Fibers	convey the n inputs $x_1, x_2, \ldots, x_n$ (or simply $\mathbf{x}$) to the unit; the n inputs $\mathbf{x}$ can originate either from outside the network or from other units in the network.

B. C. Hewitson and R. G. Crane (eds.), Neural Nets: Applications in Geography, 11–52.

Table 2.1: Elements of an artificial neural network unit.

Elements	Function
Weights	$W_1, W_2, \ldots, W_n$ (or simply **W**) that when combined with their corresponding inputs, e.g., $W_i x_i$ or W_i-x_i, determine the contribution of the input from each fiber to the activation of the unit.
Optional bias weight	W_0 that receives a constant input of one.
Activation function	u that combines the inputs and weights to produce a number, which we label $u(\mathbf{W}, W_0, \mathbf{x})$.
Transfer function	σ that operates on the unit's activation $u(\mathbf{W}, W_0, \mathbf{x})$ to produce the unit's output, which we label $o=\sigma(u(\mathbf{W}, W_0, \mathbf{x}))$.
Branching output fiber	relays the unit's output o either to specific units in the network or to objects outside of the network.

Neurons in animals apparently have a similar information flow, with one neuron's output fiber (axon) carrying activities (action potentials) to another neuron's weights (synapses). The neuron receiving the action potentials depolarizes (activation); if the depolarization is sufficiently strong, then the neuron sends action potentials to other neurons via its axon. Each step in this chain for a cortical neuron, however, can be extremely complicated (Dayhoff, 1990).

The ability of artificial neural networks to find interesting relationships in data stems from the rules for setting, or adapting, each unit's weights, the forms of the unit activation and transfer functions, and the network architecture that results from connecting a number of units according to some scheme. The properties of each unit and the network connectivity, in turn, depend strongly on the specific problem at hand. A report published by the Defense Advanced Research Projects Agency (DARPA, 1988) after a survey of the field indicated many of the research problems to which neural networks could be applied (Table 2.2). The number of different neural network models developed to address these research topics is large; therefore, in the current chapter we limit our description to some of the 'mainstream' neural networks designed to address the first four topics in the list (Figure 2.2).

In the brief discussion of the recurrent, associative, competitive, Adaptive Resonance Theory (ART) and perceptron networks (Section 2.2), we attempt to present to readers unfamiliar with neural networks some of the ideas and language necessary to understand their description and operation. The multilayer perceptron with the backpropagation of error modification rule is discussed in more detail in Section 2.2 for

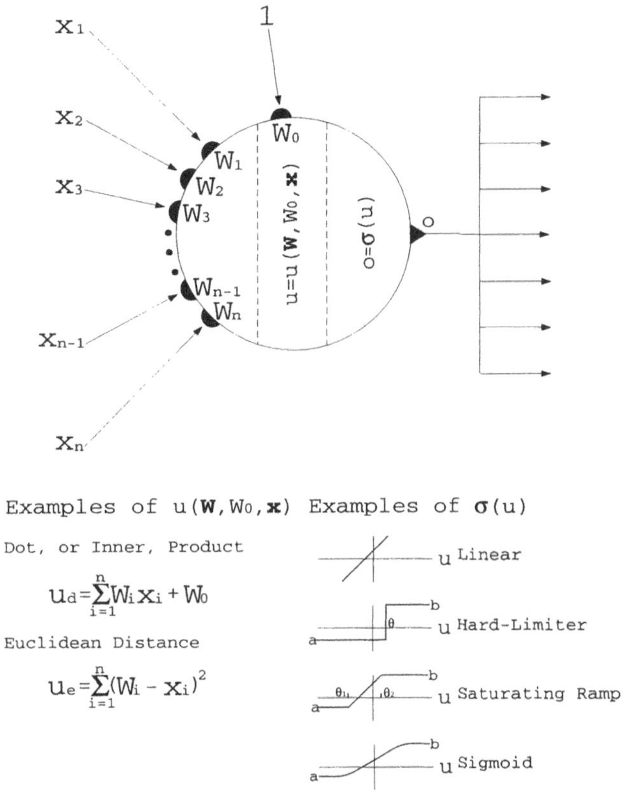

Figure 2.1: A neural network unit. The six important features of the unit are: 1) input fibers; 2) input weights; 3) bias weight; 4) activation function u; 5) transfer function σ; 6) output fibers.

two reasons. First, this algorithm is currently generating considerable interest among theoreticians, and the number of scientists and engineers using it in various applications is quite large. And second, the chapters presented in this work are based largely on the application of the backpropagation of error algorithm to data sets that are inherently complicated. We hope to provide enough detail about the algorithm to make its operation clear; two examples are presented that attempt to illustrate how the network can find useful approximations to nonlinear mappings. Section 2.3 contains a description of Kohonen's self-organizing feature map, as it is the other neural network algorithm used in the current book. This network is an unsupervised neural network, as opposed to the supervised backpropagation of error algorithm; it finds structure in the data using internal criteria for changing its weights. The descriptions of different network topologies provide

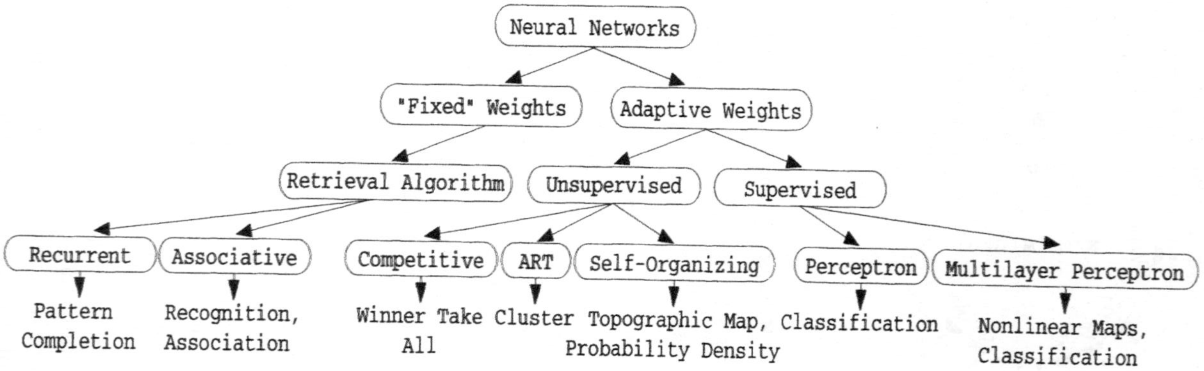

Figure 2.2: An outline of the neural networks that we discuss in the current chapter; these networks are representative of the different neural networks that have been studied in recent years.

motivation for Section 2.4, which discusses System Identification. Researchers who are studying neural networks from the perspective of *system identification* are attempting to provide some measure of validity to the solutions that are obtained from neural network approaches by establishing a closer link between ANN and statistics. We conclude with a list of current research trends in the field, such as ensembles of neural networks, chaos and neural networks, and more biologically realistic models.

Table 2.2: Applications of artificial neural networks.

Applications
Content-Addressable Memories
Pattern Completion
Pattern Recognition
Associative Memory Storage and Access
Self-Organization and Clustering
Pattern Classification
Nonlinear Mappings
Vision and Speech Preprocessing
Computational/Optimization Problems
Recognition of Time-Varying Patterns

Our aim throughout this chapter is to provide the reader enough information to understand the chapters that follow. We have attempted to provide sufficient discussion and figures so that the chapter is self-contained, in the sense that further reading should not be necessary to understand the remainder of the book. At the same time we have tried to keep things brief and simple, insofar as possible. For a more detailed description of various types of neural networks, we cannot think of better introductory books than those edited by Rumelhart et al. (1986c), McClelland et al. (1986) and Anderson and Rosenfeld (1988). The books by Rumelhart et al. (1986c) and McClelland et al. (1986) provide a fairly comprehensive outlay of the field of neural networks, while the prologues to the articles in Anderson and Rosenfeld (1988) are informative and fun to read. A number of relatively short review articles have also appeared in the literature that are readable and provide more detail about a number of networks (e.g., Lippman, 1987; McKnight, 1990; Hush and Horne, 1993). More detailed and diverse accounts of the theory and application of neural networks can also be found in the books by Hect-Nielson (1988), Dayhoff (1990), Hertz et al. (1991), Bosko (1992) and Lau (1992).

2.1 Neural Network Language and Basic Operation

2.1.1 The Basic Language

Notation in the field of neural networks is largely organized around keeping track of multiple layers of units of the type presented in Figure 2.1. The n inputs to a single unit are represented by the numbers $x_1, x_2, ..., x_n$. For simplicity, we let the vector $\mathbf{x}$ represent $x_1, x_2, ..., x_n$; that is,

$$\mathbf{x} = (x_1, \ldots, x_n) . \quad \text{1a}$$

Since $\mathbf{x}$ has n components, $\mathbf{x}$ can be represented by a point in the n-dimensional space of real numbers R^n; that is, $\mathbf{x}$ is an element of R^n, or simply $\mathbf{x} \in R^n$. For each input (x_i) to a unit there is a corresponding weight (W_i) on the unit; therefore,

$$\mathbf{W} = (W_1, \ldots, W_n, W_0) \in R^{n+1} \quad \text{1b}$$

represents all of a unit's weights, where we include the optional bias weight W_0 in $\mathbf{W}$ as well. We denote the output of the unit by o. If there are N units in a layer with n inputs per unit, then the weights of the ith unit are given by

$$\mathbf{W}_i = (W_{i1}, \ldots, W_{in}, W_{i0}) \quad \text{1c}$$

and its corresponding output by o_i. We represent all the weights in the layer by

$$\mathbf{W} = (\mathbf{W}_1, \ldots, \mathbf{W}_N) \quad \text{1d}$$

and all the unit outputs in the layer by

$$\mathbf{o} = (o_1, \ldots, o_N) . \quad \text{1e}$$

If there are multiple layers of units in the network, we use a parenthesized superscript to denote the layer. Therefore, $\mathbf{W}^{(k)}$ and $\mathbf{o}^{(k)}$ denote the weights and outputs, respectively, of the kth layer of units. In this case $\mathbf{W}$ and $\mathbf{o}$ represent all of the network weights and unit outputs, respectively. Finally, suppose we have a set of K patterns with n elements per pattern. We represent a single pattern by $\mathbf{x}^p = (x_1^p, \ldots, x_n^p)$ and all K patterns by

$$\{\mathbf{x}^1, \ldots, \mathbf{x}^K\} = \{\mathbf{x}^i | i = 1, \ldots, K\} . \quad \text{1f}$$

In this notation, $o_i^{(k),p}$ represents the output activity of the ith unit in the kth layer when the pattern $\mathbf{x}^p$ is input to the network.

For a unit to produce an output (o) from an input ($\mathbf{x} \in R^n$) that impinges on its weights ($\mathbf{W} \in R^{n+1}$), an activation function (u) and a transfer function (σ) must be defined for the unit. In Figure 2.1 we illustrate the two activation functions and the four transfer functions that are used in the networks that we describe; Table 2.3, furthermore, contains the equations that we used to produce the graphs of the transfer functions in Figure 2.1. The dot product activation function u_d is used in all the networks that we

Table 2.3: The activation and transfer functions (i.e., u and σ) used in the neural networks that we describe in this chapter.

Function Name	Mathematical Expression
Activation Functions	
Dot Product	$u_d(\mathbf{W}, W_0, \mathbf{x}) = \sum_{j=1}^{n} W_j x_j + W_0$
Euclidean Distance	$u_e(\mathbf{W}, W_0, \mathbf{x}) = \sum_{j=1}^{n} (W_j - x_j)^2$
Transfer Functions	
Linear	$\sigma_l(u) = au + b$
Hard-Limiting	$\sigma_h(u) = \begin{cases} b, \theta \le u \\ a, u < \theta \end{cases}$
Saturating Ramp	$\sigma_r(u) = \begin{cases} b, & \theta_2 < u \\ cu + d, & \theta_1 \le u \le \theta_2 \\ a, & u < \theta_1 \end{cases}$
Sigmoid	$\sigma_s(u) = \dfrac{b + ae^{-\alpha u}}{1 + e^{-\alpha u}}$

discuss, except for Kohonen's self-organizing feature map network where the Euclidean distance activation function u_e is used. Each of the four transfer functions can be found at various points in our discussion. Note that the four transfer functions are naturally separated into two groups: those transfer functions that do not contain a nonlinearity (σ_l) and those that do (σ_h, σ_r, σ_s). In the discussion of each network, we indicate which activation function and transfer function the network utilizes; we also list the parameter values for the specified transfer function. For example, suppose the units in a network incorporate the dot product activation function u_d together with the sigmoid transfer function σ_S. In this case, each unit's activation is given by the sum of the product of each input with its corresponding weight, i.e.,

$$u_d(\mathbf{W}, W_0, \mathbf{x}) = \sum_{j=1}^{n} W_j x_j + W_0. \qquad \text{1g}$$

To uniquely determine the transfer function σ_S we must specify the values of a, b and α (Table 2.3). Once a, b and α are known, the output of the unit follows from

$$o = \sigma_s(u_d) = \frac{b + ae^{-\alpha u_d}}{1 + e^{-\alpha u_d}}. \qquad \text{1h}$$

2.1.2 Recurrent Networks

Suppose, for example, that we want to identify each object in a collection of objects by its geometrical outline from an appropriate view. To make the problem interesting suppose that for each object only a part of its outline is made available with which to classify it. Therefore, the problem is to reconstruct the entire outline from only a portion of it, i.e., pattern completion. To rephrase the problem in our neural network language, $\mathbf{x} \in R^n$ represents a partial geometrical outline of one of the objects in the collection and the task is to determine the complete geometrical outline $\mathbf{y}^p \in R^n$ from which $\mathbf{x}$ is derived. As another example, let $\mathbf{y}^p \in R^n$ represent one of K possible messages that is to be sent through some medium. During the transmission of $\mathbf{y}^p$, assume noise degrades the message $\mathbf{y}^p$ to produce the received signal $\mathbf{x}$. The goal now becomes reconstructing $\mathbf{y}^p$ from $\mathbf{x}$. For simplicity, in both of the above examples assume that two constraints are placed on the prototype patterns that are to be recovered. First, the length of each prototype vector is set to one, i.e.,

$$\|\mathbf{y}^p\| = \sqrt{(y_1^p)^2 + \ldots + (y_n^p)^2} = 1. \qquad \text{2a}$$

And second, each prototype pattern is perpendicular to every other prototype pattern, i.e.,

$$\mathbf{y}^p \cdot \mathbf{y}^q = y_1^p y_1^q + \ldots + y_n^p y_n^q = 0 \qquad \text{2b}$$

for $p \neq q$. Therefore, the collection of all possible prototype patterns is given by

$$\{\mathbf{y}^i | \mathbf{y}^i \in R^n, \|\mathbf{y}^i\| = 1, \mathbf{y}^i \perp \mathbf{y}^j \ (i \neq j)\ ; i = 1, \ldots, K\} \qquad \text{2c}$$

and the task is to assign each $\mathbf{x}$ to one of the $\mathbf{y}^i$.

One avenue of attack is to use the *recurrent* (i.e., output is fed back to input) network illustrated in Figure 2.3a. In the autoassociation scheme devised by Anderson et al. (1977), the feedback weight W_{ij} to unit i from unit j is set by sums of products of the prototype pattern components:

$$W_{ij} = \sum_{p=1}^{K} c_i y_i^p y_j^p \; (i = 1, \ldots, n; j = 1, \ldots, n), \qquad 3$$

where the c_i are constants. In other words, the weights W_{ij} provide a measure of the correlation in activity between unit i and j across all prototypes in the collection. The activation function of each unit is u_d, the transfer function is σ_l with $a = 1$ and $b = 0$, and the bias weight W_{i0} is set to 0 for all i (Figure 2.1, Table 2.3). The pattern **x** is initially taken as the recurrent layer output at time 0, i.e., $\mathbf{o}(0) = \mathbf{x}$, and is fed back into the network as the input:

$$o_i(t+1) = \sigma_l(u_d(\mathbf{W}_i, 0, \mathbf{o}(t))) = \sigma_l\left(\sum_{j=1}^{n} W_{ij} o_j(t)\right) = \sum_{j=1}^{n} W_{ij} o_j(t), \qquad 4a$$

where $\mathbf{o}(t) = \mathbf{x}$ when $t = 0$. Using an eigenvalue analysis, Anderson et al. (1977) showed that the output $\mathbf{o}(t)$ tends toward linear combinations of some of the prototypes $\mathbf{y}^p$ as the iteration number t increases. Unfortunately, the scheme can be unstable and the decision of which $\mathbf{y}^p$ in the linear combination to pick as the prototype for **x** requires additional computation.

To stabilize the network and to assure positive feedback in the system, Anderson et al. (1977) changed the transfer function to σ_r, where $a = \theta_1 = -C$, $b = \theta_2 = C$, $c = 1$, and $d = 0$; all other aspects of the approach remained the same. The unit update mechanism becomes

$$o_i(t+1) = \sigma_r(u_d(\mathbf{W}_i, 0, \mathbf{o}(t))) = \sigma_r\left(\sum_{j=1}^{n} W_{ij} o_j(t)\right), \qquad 4b$$

where the unit outputs now saturate at $\pm C$ for unit activations beyond $\pm C$. The resulting model, dubbed 'brain-state-in-a-box', produces dynamics where inputs $\mathbf{x} = \mathbf{o}(0)$ ($|x_j| < C$) generally increase in magnitude until the unit outputs start to saturate; eventually, the output $\mathbf{o}(t)$ is driven to a 'corner of the box', where $o_i(t) = \pm C$ for all i. In this case, **x** is associated, or categorized, with the pattern that is encoded by the corner of the box in which the network settles.

To solve the same problem, but with binary input patterns **x** and binary prototypes $\mathbf{y}^p$ (i.e., x_j and y_j^p are either zero or one for all j and p), Hopfield (1982) also used the architecture in Figure 2.3a with a method similar to Anderson et al. (1977) for setting the weights. However, he implemented a radically different scheme for updating the unit outputs. The unit activation function is u_d, the transfer function is σ_h with $a = 0$, $b = 1$, and $\theta = 0$, and the bias weight W_{i0} is set to 0 for all i. The scheme that Hopfield implemented for updating the outputs is an *energy relaxation* technique that minimizes the function

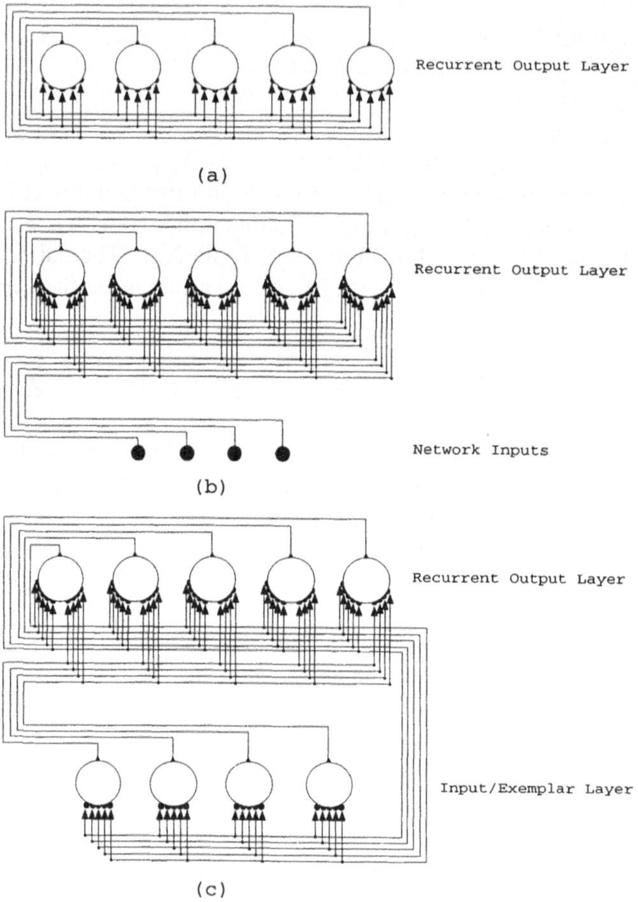

Figure 2.3: a) Recurrent network. b) An associative network if there are no recurrent connections and a competitive learning network if there are recurrent connections; if a topology is defined on the output layer, the resulting network is identical to Kohonen's self-organizing feature map. c) Adaptive Resonance Theory, or ART, network.

$$E(t) = -\frac{1}{2}\sum_{i=1}^{n}\sum_{j=1}^{n} W_{ij} o_i(t)\, o_j(t)\,. \qquad 5$$

At each iteration $t+1$, one of the units is picked at random and its output is updated according to

$$o_i(t+1) = \sigma_h\left(\sum_{j=1}^{n} W_{ij} o_j(t)\right), \qquad 6$$

where $\mathbf{o}(0) = \mathbf{x}$. Therefore, if the unit's activation is less than zero, $o_i(t+1)$ is set to zero; otherwise, $o_i(t+1)$ is set to one. The process is then iterated until no further changes occur in any of the unit outputs. Importantly, Hopfield demonstrated that this technique for updating the unit outputs is indentical to reducing the energy E at each step; that is, $\Delta E(t) \leq 0$ for all t provided the weights W_{ij} are symmetric. If the number K of prototypes is small compared to n, then the final state of the network, i.e., $\mathbf{o}(t)$ as $t \to \infty$, is generally one of the prototypes. However, as K increases in magnitude, spurious states due to *local minima* in the energy function occur and the relaxation scheme no longer consistently drives an arbitrary input pattern $\mathbf{x}$ to one of the prototypes. A number of alternative networks have been developed that perform much better than the Hopfield network as the number of stored patterns increases; one such network is the unary, or Hamming, network (see Lippmann, 1987).

In Section 2.2, we describe in full detail the generalized backward propagation of error model which is based on the idea of least-mean-square (LMS) minimization of the network response error. At this point, it is worth noting that a number of algorithms have been devised which incorporate recurrent architectures into the basic LMS framework (Williams and Zipser, 1988; Giles et al., 1990; Giles et al., 1992; Pollack, 1991). These models have been applied to a variety of time-dependent classification and predicition problems such as the identification of unknown grammars from binary strings (Giles et al., 1990; Giles et al., 1992; Pollack, 1991), nonlinear time-series prediction (Parlos et al., 1991) and speech recognition (Wang, 1993).

2.1.3 Content-Addressable Memories

The 'brain-state-in-a-box' and Hopfield networks are a form of content-addressable memory: the initial pattern of inputs ($\mathbf{x}$) can be viewed as a key to retrieve a pattern ($\mathbf{y}^p$) already stored in memory. Associative networks have a similar function. In Figure 2.3b, if n units are present in the output layer and all of the recurrent connections are set to zero, then the resulting network is an example of an associative network. The activation function for these types of networks is u_d, the transfer function is generally σ_l with $a = 1$ and $b = 0$ (e.g., Anderson, 1972; Cooper, 1973; Anderson et al., 1977), and the bias weights W_{i0} are generally set to 0 for all i; however, hard-limiting transfer functions (σ_h) together with non-zero thresholds (W_{i0}) have also been used (Amari, 1977). In associative networks the weights are generally a measure of the correlation between the jth component of the input patterns $\{\mathbf{x}^p \mid p = 1, ..., K\}$ and the ith component of the desired outputs $\{\mathbf{o}^q \mid q = 1, ..., K\}$:

$$W_{ij} = \sum_{p=1}^{K} \sum_{q=1}^{K} c_{pq} o_i^q x_j^p \; (i = 1, \ldots, n; j = 1, \ldots, n), \qquad 7$$

where the coefficients c_{pq} provide a measure of the strength of the correlation between patterns p and q. These networks have the property that if $\mathbf{x}^p$ is input to the network, then the network response is strongly dominated by $\mathbf{y}^p$ if c_{pp} is large compared to the other correlation constants and the input patterns are close to orthogonal; that is, the network 'recognizes' the input $\mathbf{x}^p$ and responds with the output pattern $\mathbf{y}^p$. However, if $c_{pq}, p \neq q$, is also significant, then $\mathbf{y}^q$ will also contribute to the output layer response to input pattern $\mathbf{x}^p$; in this case, $\mathbf{y}^q$ is said to be associated with $\mathbf{x}^p$.

An underlying motivation for the recurrent and associative network research is the form of synaptic modification proposed by Hebb (1949). Hebbian learning, as the synaptic modification rule is now called, states that the change to a synaptic weight W_{ij} is proportional to the product of the input activity x_j^p to the weight W_{ij} and the unit response o_i^p to $\mathbf{x}^p$, i.e.,

$$\Delta W_{ij} \propto o_i^p x_j^p . \tag{8}$$

Therefore, the method for setting the weights used by Anderson (1972), Cooper (1973), etc..., can be viewed as a natural consequence of using an Hebbian modification scheme to train the network during an initial period of learning.

2.1.4 *Competitive Learning and Adaptive Resonance Theory*

Grossberg investigated associative networks similar to the 'brain-state-in-a-box' network; however, fundamental to his thinking was that units should not saturate. To prevent output unit saturation, Grossberg incorporated both a sigmoid transfer function (σ_s) (Grossberg, 1973) and recurrent feedback (Grossberg, 1976) in the output layer units (Figure 2.3b). In the feedback loop, units positively reinforce their current output levels, while attempting to reduce the output of the other units in the layer. With these changes in the network architecture, the application domain of the network also shifts slightly; an arbitrary input pattern $\mathbf{x} \in R^n$ is now associated with the group of input patterns that strongly activate only one of the m output layer units. According to this kind of scheme, a pattern $\mathbf{x}$ is input to the network and the unit outputs are calculated according to

$$o_i(0) = \sigma_s\left(u_d\left(\sum_{j=1}^{n} W_{ij}^{(input)} x_j\right)\right). \tag{9a}$$

The output layer responses are subsequently used as the input to the same layer, but to the recurrent network weights:

$$o_i(t+1) = \sigma_s\left(u_d\left(\sum_{j=1}^{n} W_{ij}^{(recurrent)} o_j(t)\right)\right), \tag{9b}$$

where $W_{ij}^{(recurrent)} < 0\ (i \neq j)$, $0 < W_{ij}^{(recurrent)}$, and the weights $W_{ij}^{(recurrent)}$ are generally equal in magnitude for $i \neq j$. (Note that this is not the exact equation used in Grossberg (1976)). The recurrent feedback turns off all of the output units, except for the one output unit that is driven most effectively by the input pattern **x**; pattern **x** is said to belong to the group represented by the one remaining active output unit.

In an extension of the competitive layer network in Figure 2.3b, Carpenter and Grossberg (1987) developed the algorithm now known as adaptive resonance theory (ART; Figure 2.3c). In this algorithm the network input patterns are imposed as activities of actual input layer units. The algorithm works just as the competitive layer network does to produce one active output unit. The activity of the active output layer unit is then fed back to the input layer units and a new pattern of activity is induced in these units. If the pattern of activity induced in the input layer units is similar to the original pattern **x**, then **x** is categorized in the group represented by the active output unit. If the activity induced in the input layer units is not similar to **x**, then **x** is assumed to come from a new category. Consequently, an additional output layer unit is created that both remains active (after competitive inhibition) when **x** is input to the network and induces the activities **x** in the input layer units. In other words, **x** becomes the exemplar of a new group represented by the new output unit. Note that ART is an *unsupervised* neural network; there is no information provided to the network that tells the network how to respond to specific input patterns. Rather, an internal measure of similarity between an input pattern of activity and the activity induced in the input layer units from output layer feedback provides information about the pattern; the network adjusts accordingly.

2.1.5 The Perceptron

Consider the network in Figure 2.3b with a single output unit and no recurrent connections. The resulting network consists of one unit with n input fibers conveying the activities **x** (i.e., Figure 2.1). If the activation function is u_d ($W_0 \neq 0$) and the transfer function is σ_h ($a = 0, b = 1, \theta = 0$), the resulting unit is similar to a *perceptron* (Rosenblatt, 1958; Block, 1962). The importance of the perceptron stems from the fact that it can split a space into two half-spaces using a hyperplane whose orientation and location are set by the weights **W**. For example, consider the two distinct groups of points in R^2 illustrated in Figure 2.4 and a unit with two input weights W_1 and W_2, as well as a bias weight W_0. If $W_1 = 1$, $W_2 = 1$, and $W_0 = -1$, then the unit output is 1 if (x, y) is to the right of the line in Figure 2.4 and 0 otherwise. In general, the location of the hyperplane of a perceptron unit can always be determined from the equation

$$\sum_{j=1}^{n} W_j x_j + W_0 = 0. \qquad \text{10a}$$

In the case discussed above we have

$$W_1 x_1 + W_2 x_2 + W_0 = 0; \qquad \text{10b}$$

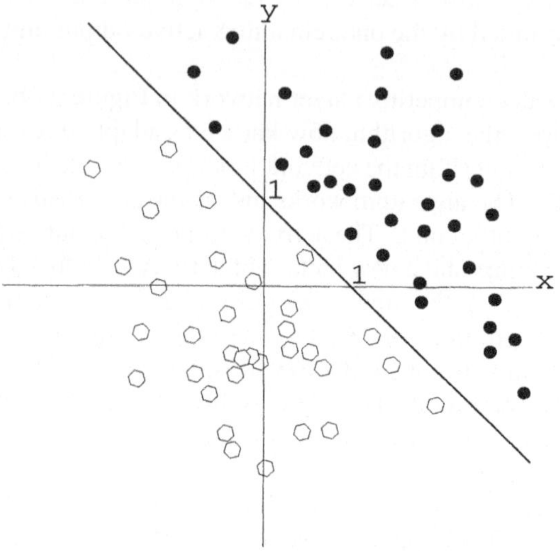

Figure 2.4: The one-dimensional hyperplane in R^2 for a unit with the two input weights $W_1 = 1$ and $W_2 = -1$ and a bias weight of $W_0 = -1$. The unit's hyperplane divides R^2 into two regions since it responds positively to the right of the hyperplane (filled circles) and negatively to the left of the hyperplane (open hexagons).

therefore,

$$x_2 = -\left(\frac{W_1}{W_2}\right)x_1 - W_0. \qquad 10c$$

For $W_1 = W_2 = 1$ and $W_0 = -1$, we have $x_2 = -x_1 + 1$, which is the equation of the line illustrated in Figure 2.4.

Much of the early excitement about perceptrons resulted from their 'built-in' ability to adapt the weights **W** in order to map a set of example patterns $\mathbf{x}^p$ to their desired target values $\mathbf{y}^p$. If the training examples could be separated by a line (e.g., Figure 2.4), then the 'perceptron convergence' learning rule guaranteed that the weights would be found that produce the line.

Another important learning rule that was applied to 'perceptron-like' units, called the delta rule or least-mean-square (LMS) rule, was developed by Woodrow and Hoff (1960). The delta modification rule is a *supervised* scheme whereby the unit weights are adapted from the input patterns $\mathbf{x}^p$, their respective target value y^p (called the input pattern label), and the actual unit output o^p:

$$\Delta W_j = -\eta\,(o^p - y^p)\,x_j^p\;(j = 1, \ldots, n). \qquad 11$$

Supervision means that the network has information about what the output of the network should be to each of the patterns in the training set. Patterns are iteratively input to the unit and the weights changed according to (11) until the weights stabilize or no longer change by significant amounts. A generalization of this straightforward learning rule is the heart of the backpropagation of error learning algorithm, which we will now discuss.

2.2 Multilayer Perceptrons and the Backpropagation of Error Algorithm

2.2.1 Neural Networks, Nonlinear Mappings and Completeness

In the context of nonlinear mappings, one might ask if a neural network map can exactly reproduce any continuous mapping f from I^n to R, where $I^n = [0,1]^n$ is the unit cube in R^n. This question naturally arises as a result of a theorem by Kolmogorov (1957), which is stated by Lorentz (1976) as: "There exist fixed continuous increasing functions $\phi_{pq}(x)$, on $I = [0,1]$ so that each continuous function f on I^n can be written in the form

$$f(x_1, \ldots, x_n) = \sum_{q=1}^{2n+1} g_q\left(\sum_{p=1}^{n} \phi_{pq}(x_p)\right), \qquad 12$$

where g_q are properly chosen continuous functions of one variable." According to this theorem, a network topology as illustrated in Figure 2.5 can implement the function $f(x_1,\ldots, x_n)$ exactly. However, as Girosi and Poggio (1989), among others, state, the theorem by Kolmogorov is not directly applicable to neural network maps. First, in neural network maps each unit in the network implements the same transfer function σ and hence there are no units in the network that can implement the functions g_q whose functional form depends on the function f being reproduced. And second, the functions $\phi_{pq}(x_p)$, and hence g_q, are not necessarily smooth. Since some degree of smoothness is generally required in the unit transfer functions for a neural network to learn and generalize, they cannot implement all of the possible functional dependencies necessary in Kolmogorov's theorem. Consequently, neural networks cannot exactly reproduce every continuous function f from I^n to R according to Kolmogorov's prescription.

Networks composed of units with identical transfer functions can, however, *approximate* continuous functions f from I^n to R to any required precision. Carroll and Dickinson (1989), Cybenko (1989), Funahashi (1989), Hornik et al. (1989), and Jones (1989) have demonstrated that for each continuous function $f: I^n \rightarrow R$ and $\varepsilon > 0$, there is a function $G(\mathbf{x})$ of the form

$$G(\mathbf{x}) = \sum_{i=1}^{N^{(1)}} W_{1i}^{(2)} \sigma\left(\sum_{j=1}^{n} W_{ij}^{(1)} x_j + \theta_i\right) \qquad 13a$$

such that

$$|G(\mathbf{x}) - f(\mathbf{x})| < \varepsilon \qquad 13b$$

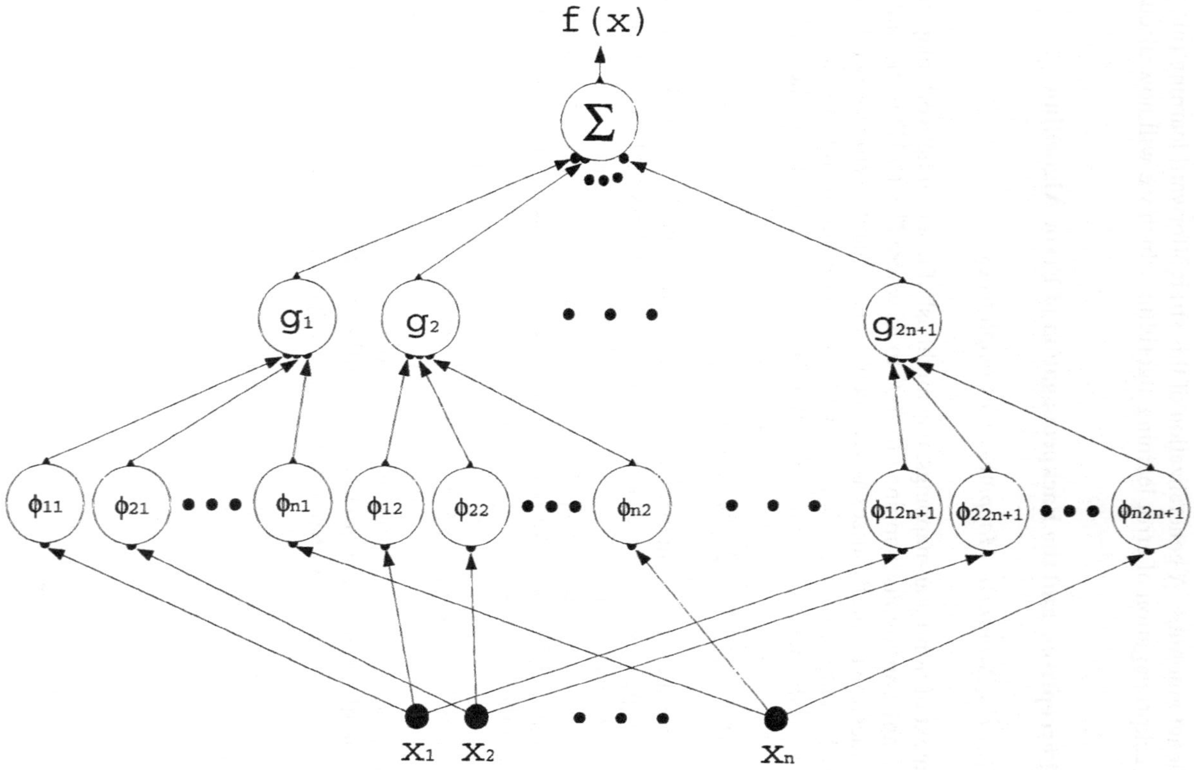

Figure 2.5: The topology of the network based on Kolomogorov's theorem. Note that the same transfer function is not used in each of the network units, hence the network is not considered to be a neural network.

for all $\mathbf{x} \in I^n$. The constraints on σ change slightly from one proof to the next. All of the proofs are valid for functions σ that are continuous, monotonic, and sigmoidal, where a function is called sigmoidal if it asymptotes at $-\infty$ and ∞, e.g.,

$$\sigma(x) \rightarrow \left(\begin{array}{l} b, as \;\; x \rightarrow \infty \\ a, as \;\; x \rightarrow -\infty \end{array} \right. \qquad 14$$

where $a < b$ (Table 2.3). Some of the proofs are true for more general forms of σ. For example, the results of Cybenko (1989) and Jones (1989) also apply to sigmoidal functions that are neither continuous nor monotonic. Stinchcombe and White (1989) conclude that the exact form of the transfer function in the units is not crucial to the success of the network; rather, the number of units in the hidden layer, which we will define shortly, is crucial since it determines the network's ability to approximate an arbitrary continuous mapping f from I^n to R. Similar results also apply to networks with more than two layers of units (e.g., Cybenko, 1988; Moore and Pogio, 1988; Funahashi, 1989; Hornik et al., 1989) and to networks with more than one output unit (e.g., Funahashi, 1989; Hornik et al., 1989).

Therefore, for a continuous mapping f from R^n to R^m, i.e.,

$$f(\mathbf{x}) = (f^1(\mathbf{x}), \ldots, f^m(\mathbf{x})) \qquad 15a$$

where the f^i ($i = 1, \ldots, m$) are functions from R^n to R, there are mappings

$$G_1(\mathbf{x}) = (G_1^1(\mathbf{x}), \ldots, G_1^m(\mathbf{x})) \qquad 15b$$

and

$$G_2(\mathbf{x}) = (G_2^1(\mathbf{x}), \ldots, G_2^m(\mathbf{x})), \qquad 15c$$

where

$$G_1^i(\mathbf{x}) = \sum_{j=1}^{N^{(1)}} W_{ij}^{(2)} \sigma\left(\sum_{k=1}^{n} W_{jk}^{(1)} x_k + \theta_j \right) \qquad 15d$$

and

$$G_2^i(\mathbf{x}) = \sum_{j=1}^{N^{(2)}} W_{ij}^{(3)} \sigma\left(\sum_{k=1}^{N^{(1)}} W_{jk}^{(2)} \sigma\left(\sum_{l=1}^{n} W_{kl}^{(1)} x_l + \theta_k^{(1)} \right) + \theta_j^{(2)} \right), \qquad 15e$$

that approximate the mapping f to any arbitrary precision.

As an example, consider the function $f(x) = 4x(1 - x)$ (Figure 2.6a). We can write $f(x)$ as the difference of two strictly increasing functions $g(x)$ and $h(x)$ (Figure 2.6b):

$$f(x) = g(x) - h(x), \qquad 16a$$

where

$$g(x) = \begin{cases} 4x(1-x), & 0 \le x \le 1/2 \\ 1, & 1/2 < x \le 1 \end{cases} \quad \text{16b}$$

and

$$(x) = \begin{cases} 0, & 0 \le x \le 1/2 \\ 1-4x(1-x), & 1/2 < x \le 1 \end{cases} . \quad \text{16c}$$

According to the constructive proof of Jones (1990), we can approximate $g(x)$ and $h(x)$, and hence $f(x)$, to within ε for all x by the sums of sigmoids

$$G(x) = \frac{2}{l} \sum_{i=1}^{\frac{l}{2}} \sigma_h (x - a_{2i-1}) \quad \text{17a}$$

and

$$H(x) = \frac{2}{l} \sum_{i=1}^{\frac{l}{2}} \sigma_h (x - b_{2i-1}) , \quad \text{17b}$$

respectively, where

$$a_i = g^{-1} \left(\frac{i}{l}\right) , \quad \text{17c}$$

$$b_i = h^{-1} \left(\frac{i}{l}\right) , \quad \text{17d}$$

and l is the smallest even integer greater than or equal to $1/\varepsilon$. For the hard-limiting transfer function σ_h, $a = 0$, $b = 1$, and $\theta = 0$. Therefore,

$$G_1(x) = G(x) - H(x) = \sum_{j=1}^{l} W_j \sigma_h (x + \theta_j) , \quad \text{18a}$$

where

$$W_j = \begin{cases} \frac{2}{l}, & j \le \frac{l}{2} \\ -\frac{2}{l}, & \frac{l}{2} < j \end{cases} \quad \text{18b}$$

and

$$\theta_j = \begin{cases} -a_{2j-1}, & j \le \frac{l}{2} \\ -b_{2j-1-l}, & \frac{l}{2} < j , \end{cases} \quad \text{18c}$$

approximates $f(x)$ to within ε for all x.

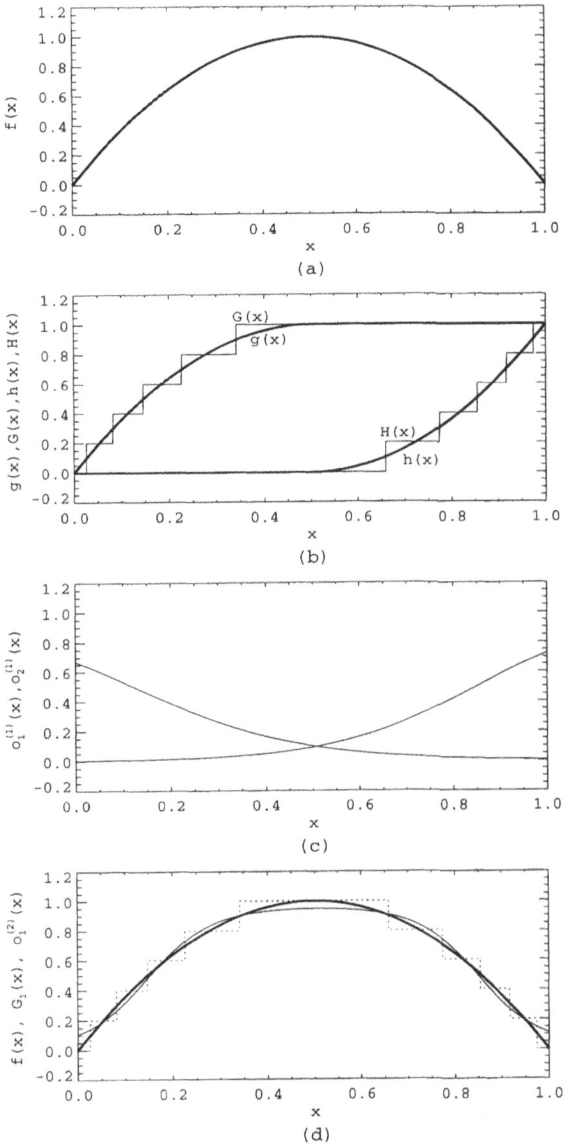

Figure 2.6: Approximations of the function $f(x)$ using Jones (1990) constructive technique and one hidden layer multilayer perceptron trained with the backpropagation of error algorithm. a) Th function $f(x) = 4x(1-x)$. b) $f(x) = g(x) - h(x)$, where $g(x)$ and $h(x)$ are non-decreasing functions $G(x)$ and $H(x)$ are approximations to $g(x)$ and $h(x)$, respectively, such that $| G(x) - g(x) | < 0.1$ an $| H(x) - h(x) | < 0.1$ for all x; therefore, $| f(x) - (G(x) - H(x)) | < 0.1$ for all x. c) The outputs $o_1^{(1)}$ and $o_2^{(1)}$ of the two units in the hidden layer after training with the backpropagation of error modification rule. d) $f(x)$ (thick solid line), the multilayer perceptron approximation of $f(x)$, $o_1^{(2)}$ (thin solid line), and Jones (1990) constructive approximation of $f(x)$, $G(x)$-$H(x)$, (dashed line).

In Figure 2.6b we illustrate $h(x)$, $g(x)$, $H(x)$, and $G(x)$ for $\varepsilon = 0.1$. Now consider $G(x)$. First, note the $l = 1/0.1 = 10$. Using the quadratic equation, we can write the inverse of $g(x)$ as

$$x = g^{-1}\left[\frac{i}{l}\right] = 0.5\left[1 - \sqrt{1 - \frac{i}{l}}\right]. \qquad \text{18d}$$

Therefore, $a_1 = 0.026$, $a_3 = 0.082$, $a_5 = 0.146$, $a_7 = 0.226$, and $a_9 = 0.342$. $G(x)$ becomes

$$G(x) = 0.2 \sum_{i=1}^{5} \sigma_h (x - a_{2i-1}) \cdot \qquad \text{19}$$

Since the hard-limiting transfer function is used in the construction, each sigmoid in the sum for $G(x)$ is 0 to the left of its corresponding threshold a_i and 1 to the right of it. Therefore, $G(x)$ is 0 for $x < a_1$; as x increases, $G(x)$ increments by $2/l$ (i.e., 0.2) each time x increases beyond one of the thresholds $a_1 < a_3 < a_5 < a_7 < a_9$. Therefore, $G(x)$ approximates $g(x)$ by a step function with a sufficient number of appropriately located steps to stay within ε of $g(x)$ for all x. Note that if $g(x)$ were not smooth, the number of steps, i.e., sigmoids, necessary to approximate it to some given accuracy ε may become extremely large. For really nasty functions $g(x)$, the number of units could go to infinity, e.g., a function $g(x)$ with an infinite number of discontinuities that can be ordered. The same considerations hold for $H(x)$ as well.

Another example is the mapping $f: I^2 \rightarrow R$ illustrated in Figure 2.7. For points (x, y) in the black area $f(x, y) = 0$ and for points (x, y) in the white area $f(x, y) = 1$.

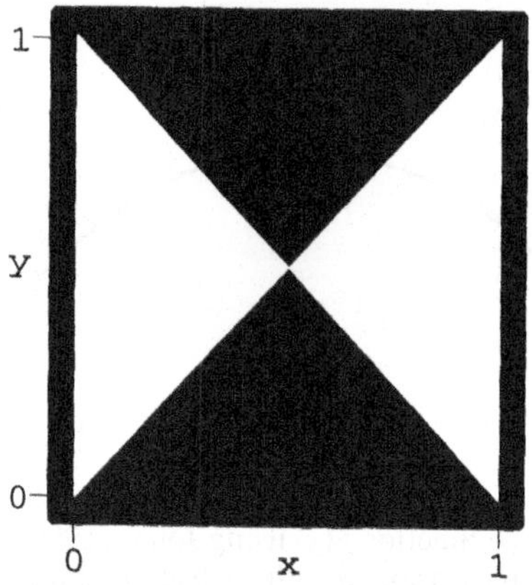

Figure 2.7: The function $f(x, y)$ on the unit cube in R^2. $f(x, y) = 0$ in the black areas and $f(x, y) = 1$ in the white areas.

Gibson and Cowan (1990) have shown that the mapping f cannot be exactly reproduced using the mapping $G_1(\mathbf{x})$ with two layers; nevertheless, we know the function f can be approximated to any arbitrary precision with a two layer network $G_1(\mathbf{x})$. To illustrate how a sum of sigmoids could approximate f, we again use Jones' (1990) constructive approximation scheme. For two dimensions, however, the algorithm is slightly more complicated, as the orientation and magnitude of each sigmoid's hyperplane and bias, respectively, must be determined. The algorithm that we implemented is based on the work of Jones (1992). We start the algorithm by considering one unit's response (o_1) to the inputs x and y:

$$o_1 = o_1(x, y) = c_1\sigma_s(W_{11}x + W_{12}y + W_{10}) . \tag{20a}$$

W_{11} and W_{12} are the unit's two weights that receive the input x and y, respectively, while W_{10} is the bias weight; c_1 is simply a scale factor. Using an appropriate search technique, c_1, W_{11}, W_{12}, and W_{10} are adjusted until the LMS error between the original function $f(x,y)$ and the unit map $o_1(x, y)$, i.e.,

$$E_1 = \sum_{i=0}^{20}\sum_{j=0}^{20}\left[f\left(\frac{i}{20}, \frac{j}{20}\right) - o_1\left(\frac{i}{20}, \frac{j}{20}\right)\right]^2 , \tag{20b}$$

is minimized. A new function is created by taking the difference between $f(x, y)$ and $o_1(x,y)$:

$$f_{err}^1(x, y) = f(x, y) - o_1(x, y) . \tag{20c}$$

A second unit with parameters c_2, W_{21}, W_{22}, and W_{20} is introduced into the scheme and its parameters are subsequently adjusted using some search technique until the error

$$E_2 = \sum_{i=0}^{20}\sum_{j=0}^{20}\left[f_{err}^1\left(\frac{i}{20}, \frac{j}{20}\right) - o_2\left(\frac{i}{20}, \frac{j}{20}\right)\right]^2 \tag{20d}$$

is minimized, where

$$o_2 = o_2(x, y) = c_2\sigma_s(W_{21}x + W_{22}y + W_{20}) . \tag{20e}$$

Now we can write E_2 as

$$E_2 = \sum_{i=0}^{20}\sum_{j=0}^{20}\left[f\left(\frac{i}{20}, \frac{j}{20}\right) - f_2\left(\frac{i}{20}, \frac{j}{20}\right)\right]^2 , \tag{20f}$$

where

$$f_2(x, y) = c_1\sigma_s(W_{11}x + W_{12}y + W_{10}) + c_2\sigma_s(W_{21}x + W_{22}y + W_{20}) . \tag{20g}$$

Therefore, each time the process is iterated, the LMS error decreases and the number of units involved in the solution increases by one. Therefore, after n iterations

$$E_n = \sum_{i=0}^{20} \sum_{j=0}^{20} \left[f_{err}^{n-1}\left(\frac{i}{20}, \frac{j}{20}\right) - f_n\left(\frac{i}{20}, \frac{j}{20}\right) \right]^2, \qquad \text{20h}$$

where $f_n(x, y)$ is the sum of the outputs of n units.

We applied this scheme to the Figure 2.7 data, performing ten iterations. In the left column of Figure 2.8 we graph the resulting $f_i(x, y)$ for $i = 1, 2, 3, 4$, and 10. Inspection of the orientation and location of the level contours provides an indication of the direction of the vector (W_{n1}, W_{n2}) and the value of the bias W_{n0} for each sigmoid. For example, in the first row there is one sigmoid unit with a bias of about $W_0 \approx -0.9$ and weights $(W_1, W_2) \approx (0.0, 1.0)$; therefore, the sigmoid hyperplane is parallel to the x-axis and is located at $y \approx 0.9$.

2.2.2 The Backpropagation of Error Neural Network Algorithm

The multilayer perceptron with the backpropagation of error technique for adjusting the weights is another technique for approximating functions from R^n to R^m. In fact, the network representations of $G_1(\mathbf{x})$ and $G_2(\mathbf{x})$ defined in the previous section are identical to the multilayer perceptron architecture with one and two hidden layers, respectively. For example, the multilayer perceptron architecture with one hidden layer illustrated in Figure 2.9a is an example of $G_1(\mathbf{x})$ from $R^4 \rightarrow R^2$ with $N^{(1)} = 6$, while the multilayer perceptron with two hidden layers illustrated in Figure 2.9b is an example of $G_2(\mathbf{x})$ from $R^4 \rightarrow R^3$ with $N^{(1)} = N^{(2)} = 6$.

Now in real-world applications one does not know the form of the function being approximated, and one generally has only a limited number of samples from it. Let $\{(\mathbf{x}^p, \mathbf{y}^p) \mid \mathbf{x}^p \in R^n, \mathbf{y}^p \in R^m;\ p = 1, \ldots, K\}$ represent the K available samples of input data $\mathbf{x}^p$ (independent variables) with their corresponding output values $\mathbf{y}^p$ (dependent variables). For the multilayer perceptron network to be useful, it must incorporate an algorithm that can use the K data samples to find the number of hidden units $N^{(1)}$ (and $N^{(2)}$ if there are two hidden layers) and the values of the weights $W_{lm}^{(k)}$ such that if $\mathbf{x}^p$ is input to the network then the network output is $\mathbf{y}^p$. Furthermore, the mapping that the network develops should perform satisfactorily on novel data as well, i.e., the network should be able to generalize from the K training samples. We now describe the backpropagation of error algorithm (Rumelhart et al., 1986a and 1986b), and then we use it to produce approximations for the two functions illustrated in Figures 2.6a and 2.7.

The first step in the algorithm is to calculate the network output to the input pattern $\mathbf{x}^p$. To start, $\mathbf{x}^p \in R^n$ is input to the first, and perhaps only, hidden layer of units. The

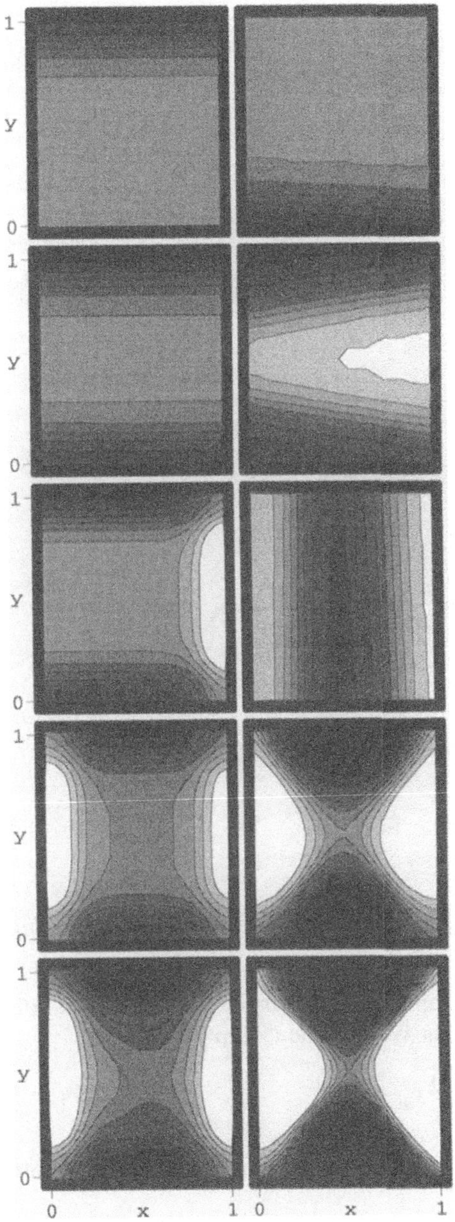

Figure 2.8: Approximations to $f(x, y)$ using Jones (1990, 1992) constructive technique (left column) and the multilayer perceptron with one hidden layer trained with the backpropagation of error rule (right column). Rows 1, 2, 3, 4, and 5 correspond to approximations using 1, 2, 3, 4 , and 10 units, respectively.

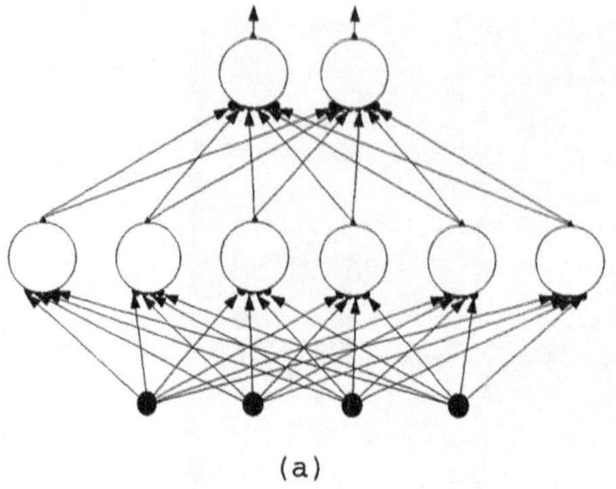

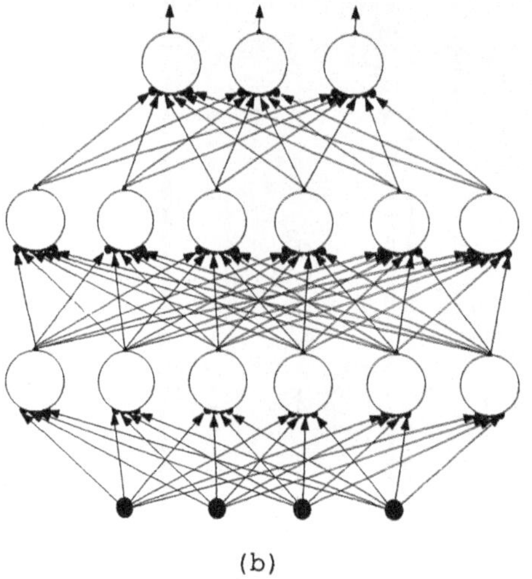

Figure 2.9: Multilayer perceptron with one (a) and two (b) hidden layers.

activation of the lth unit of the first hidden layer, i.e., $u_{d,l}^{(1),p}$, is found by taking the dot product of $\mathbf{x}^p$ and the weights $\mathbf{W}_l^{(1)}$ to the lth unit:

$$u_{d,l}^{(1),p} = \sum_{m=1}^{n} W_{lm}^{(1)} x_m^p + W_{l0}^{(1)}. \qquad \text{21a}$$

The resulting activations for each hidden layer unit are passed through the sigmoid, i.e.,

$$o_l^{(1),p} = \sigma_s(u_{d,l}^{(1),p}) = \frac{1}{1+e^{-\alpha u_{d,l}^{(1),p}}}, \qquad \text{21b}$$

and the resulting values are used in a dot product with the weights of each unit in the next layer, which is either the output layer or the second hidden layer:

$$u_{d,l}^{(2),p} = \sum_{m=1}^{N^{(1)}} W_{lm}^{(2)} o_m^{(1),p} + W_{l0}^{(2)} . \tag{22a}$$

If the second layer is the output layer, the resulting $u_{d,l}^{(2),p}$ can be taken as the network outputs or they can be passed through the sigmoid, i.e.,

$$o_l^{(2),p} = \sigma_s(u_{d,l}^{(2),p}) = \frac{1}{1+e^{-\alpha u_{d,l}^{(2),p}}}, \tag{22b}$$

in which case the values $o_l^{(2),p}$ are taken as the network outputs. If $o_l^{(2),p}$ is the second hidden layer output, then the process is repeated again to produce the activations $u_{d,l}^{(3),p}$ in the output layer units; again, the network outputs can be taken as either $u_{d,l}^{(3),p}$ or $o_l^{(3),p}$. In the following discussion we will assume that the network outputs are $o_l^{(k),p}$ $(k = 2, 3)$.

Since backpropagation of error uses a supervised learning technique that minimizes an energy function E_p, the network output $\mathbf{o}^{(k),p}$ to the pattern $\mathbf{x}^p$ is compared to the target output $\mathbf{y}^p$ (i.e., supervision) and the resulting information is used to change the network weights $\mathbf{W}$ so that the magnitude of the error E_p is smaller the next time $\mathbf{x}^p$ is input to the network. The error function E_p in the algorithm is a measure of the Euclidean distance between $\mathbf{y}^p$ and $\mathbf{o}^{(k),p}$:

$$E_p = \frac{1}{2}\sum_{i=1}^{N^{(k)}} (y_i^p - o_i^{(k),p})^2 . \tag{23}$$

To illustrate how the modification rule works, first consider the mth unit in the kth layer. We know that

$$o_m^{(k),p} = \sigma_s(u_{d,m}^{(k),p}) = (1+e^{-\alpha u_{d,m}^{(k),p}})^{-1}, \tag{24a}$$

and

$$u_{d,m}^{(k),p} = \sum_{l=1}^{N^{(k-1)}} W_{ml}^{(k)} o_l^{(k-1),p} + W_{m0}^{(k)} . \tag{24b}$$

Based on these equations, there are four important derivatives that are used repeatedly in the modification rule. First,

$$\frac{\partial o_m^{(k),p}}{\partial u_{d,m}^{(k),p}} = \frac{\partial \sigma_s(u_{d,m}^{(k),p})}{\partial u_{d,m}^{(k),p}} = -(1+e^{-\alpha u_{d,m}^{(k),p}})^{-2}(-\alpha e^{-\alpha u_{d,m}^{(k),p}}) . \tag{25}$$

Solving for $e^{-\alpha u_{d,m}^{(k),p}}$ in (24a) yields

$$e^{-\alpha u_{d,m}^{(k),p}} = \frac{1 - o_m^{(k),p}}{o_m^{(k),p}} ; \qquad 26$$

therefore, (25) can be written as

$$\frac{\partial o_m^{(k),p}}{\partial u_{d,m}^{(k),p}} = \alpha o_m^{(k),p} (1 - o_m^{(k),p}) . \qquad 27$$

The second and third important derivatives are

$$\frac{\partial u_{d,i}^{(k),p}}{\partial W_{ml}^{(k)}} = \delta_{i,m}^{Kron} o_l^{(k-1),p} , \qquad 28a$$

where $\delta_{i,m}^{Kron}$ is the Kronecker delta, or identity matrix:

$$\delta_{i,m}^{Kron} = \begin{cases} 0, & for\ i \neq m \\ 1, & for\ i = m \end{cases} \qquad 28b$$

and

$$\frac{\partial u_{d,m}^{(k),p}}{\partial o_l^{(k-1),p}} = W_{ml}^{(k)} . \qquad 29$$

In addition to these derivatives, the last and perhaps most important of the four derivatives, which is defined in terms of the above derivatives, is the error signal. For the ith cell in the kth layer of connections, it is defined to be:

$$\delta_i^{(k),p} = -\frac{\partial E_p}{\partial u_{d,i}^{(k),p}} . \qquad 30$$

Now the generalized delta rule used in backpropagation of error is

$$\Delta W_{ml}^{(k)} = -\eta \frac{\partial E_p}{\partial W_{ml}^{(k)}} \qquad 31$$

for the lth weight to the mth unit of the kth layer. For an output unit weight ($W_{ml}^{(k)}$), we have

$$\frac{\partial E_p}{\partial W_{ml}^{(k)}} = \sum_{i=1}^{N^{(k)}} \frac{\partial E_p}{\partial o_i^{(k),p}} \frac{\partial o_i^{(k),p}}{\partial u_{d,i}^{(k),p}} \frac{\partial u_{d,i}^{(k),p}}{\partial W_{ml}^{(k)}} \qquad 32a$$

or

$$\frac{\partial E_p}{\partial W_{ml}^{(k)}} = \frac{1}{2} \sum_{i=1}^{N^{(k)}} 2 (y_i^p - o_i^{(k),p}) \left(-\frac{\partial o_i^{(k),p}}{\partial u_{d,i}^{(k),p}} \right) \left(\frac{\partial u_{d,i}^{(k),p}}{\partial W_{ml}^{(k)}} \right) . \qquad 32b$$

Note that this may be rewritten in terms of the error signal using the fact that:

$$\delta_i^{(k),p} = -\eta \frac{\partial E_p}{\partial u_{d,i}^{(k),p}} = -\eta \frac{\partial E_p}{\partial o_i^{(k),p}} \frac{\partial o_i^{(k),p}}{\partial u_{d,i}^{(k),p}} \quad . \tag{33}$$

Using (27), (28a), (32) and (33), we have

$$\Delta W_{ml}^{(k)} = -\eta \frac{\partial E_p}{\partial W_{ml}^{(k)}} = \eta \sum_{i=1}^{N^{(k)}} \delta_i^{(k),p} \delta_{i,m}^{Kron} o_l^{(k-1),p} = \eta \delta_m^{(k),p} o_l^{(k-1),p} \quad . \tag{34a}$$

In a similar fashion, we can show that

$$\Delta W_{ml}^{(k-1)} = \left(-\eta \frac{\partial o_m^{(k-1),p}}{\partial u_{d,m}^{(k-1),p}} \right) \sum_{i=1}^{N^{(k)}} \delta_i^{(k),p} W_{im}^{(k)} o_l^{(k-2),p} = \eta \delta_m^{(k-1),p} o_l^{(k-2),p} \tag{34b}$$

for the hidden layer unit weights to the layer of units preceding the output layer. Note that (34a) and (34b) can be summarized as:

$$\Delta W_{ml}^{(k)} = \eta \delta_m^{(k),p} o_l^{(k-1),p} \, , \tag{35}$$

for all layers k, but with the error signal defined to be:

$$\delta_m^{(k),p} = \begin{cases} (y_m^p - o_m^{(k),p}) \, \alpha o_m^{(k),p} (1 - o_m^{(k),p}) & (output\ layer) \\ \alpha o_m^{(k),p} (1 - o_m^{(k),p}) \displaystyle\sum_{i=1}^{N^{(k+1)}} \delta_i^{(k+1),p} W_{im}^{(k+1)} & (hidden\ layer) \end{cases} \quad . \tag{36}$$

Note that to calculate the change in the weights in layer k - 1 one must use the error terms $\delta_i^{(k),p}$; that is, the errors must be propagated from the output units back towards the input layer units, hence the name backpropagation of error for this modification algorithm. This recursive calculation extends backward through the network if there are more than two layers of weights, allowing a simple and efficient calculation of the weight modifications. Now $\Delta W_{ml}^{(k)}$ represents the change in the weight $W_{ml}^{(k)}$ when pattern $\mathbf{x}^p$ is input to the network. During a training session of the network, the different training set samples ($\mathbf{x}^p$, $\mathbf{y}^p$) are iteratively input to the network and the weights adjusted until the network weights stabilize or some criterion for cessation of training is satisfied, such as performance on a cross-validatory data set.

To provide some examples of the backpropagation of error algorithm, we applied it to the functions illustrated in Figures 2.6a and 2.7. For the function in Figure 2.6a the training set consisted of the inputs $x^i = i/100$ and their corresponding target values $y^i = 4(i/100)(1 - i/100)$ ($i = 1, ..., 100$). The network architecture is similar to Figure 2.9a, but with a single network input (x), a single unit output (y), and a hidden layer with anywhere from one to ten units. The activation function was u_d ($W_0 \neq 0$) and the transfer function was σ_s ($a = 0, b = 1, \alpha = 20$). In Figures 2.6c and 2.6d we illustrate the results of the algorithm when there were two units in the hidden layer. Inspection of the two unit outputs in the hidden layer (Figure 2.6c) immediately indicates two things. There is no way that a single hidden layer with only one unit can adequately model $f(x)$. A single unit would be able to model $f(x)$ fairly reasonably for either $0 \leq x \leq 0.5$ or $0.5 \leq x \leq 1.0$; however, a single unit does not have the degrees of freedom to model the peak in $f(x)$ at $x = 0.5$. In contrast, two units in the hidden layer are sufficient to produce a fairly good approximation to $f(x)$, with each unit exclusively modelling $f(x)$ on the domain from $0 \leq x \leq 0.5$ and $0.5 \leq x \leq 1.0$ (Figure 2.6c). In fact, two units in the hidden layer produced such a good fit to the data (Figure 2.6d) that using up to ten units in the hidden layer did not dramatically improve algorithm performance for a 'reasonable' number of training iterations.

For the function in Figure 2.7 the training set consisted of 441 input-output data pairs. The inputs were given by $\mathbf{x}^i = (i/20, j/20)$ and their corresponding target values by $\mathbf{y}^i = f(i/20, j/20)$ $(i = 0, ..., 20; j = 0, ..., 20)$, where

$$f(\frac{i}{20}, \frac{j}{20}) = \begin{cases} 0, & i/20 \leq 0.5 \quad \text{and either} \quad j < i \quad \text{or} \quad 20 - i < j \\ 0, & 0.5 < i/20 \quad \text{and either} \quad j < 20 - i \quad \text{or} \quad i < j \\ 1, & \text{otherwise}. \end{cases} \qquad 37$$

The inputs correspond to points uniformly distributed over the unit cube in Figure 2.7; the output value is zero if the input is within a black region and one if the input is within a white region. The network architecture is again similar to Figure 2.9a, but with two network inputs (x, y), a single unit output (o), and a hidden layer with anywhere from one to ten units. The activation function was $u_d(W_0 \neq 0)$ and the transfer function was $\sigma_s(a = 0, b = 1, \alpha = 20)$. In the right hand column of Figure 2.8 we illustrate the results of the algorithm when there were 1, 2, 3, 4, and 10 units in the hidden layer. Unlike the results in the left hand column of the figure, the rows in the right hand column do not form a natural progression. Each time the number of hidden units is changed we have to start from scratch and retrain the neural network. Since the initial weights in the network are random values between -0.5 and 0.5, there is no guarantee that two consecutive runs with the same architecture will produce the same results. Consequently, a solution found with $N^{(1)}$ units will generally not have a bearing on the solution found with $N^{(1)} + 1$ units. Such a discrepancy is readily apparent in a comparison of rows two and three. From inspection of Figure 2.8 we conclude that at least four hidden units are needed in order to 'capture' the primary 'criss-cross' structure in the data and going to ten hidden units does slightly improve the model of the data.

In the above two examples of the backpropagation algorithm there are a number of issues that we glossed over. These issues pertain to: the shape of the energy surface, the neural network architecture and size, and generalization. For a gradient descent algorithm like backpropagation of error, the shape of the underlying energy surface strongly affects the quality of the solution mapping. If the energy surface is complex, the descent in energy may drive the network to local minima where the solution mappings are poor. In the above two examples we attempted to avoid local minima by using the standard technique of repeatedly training the neural network from scratch using different randomized weights to initialize it; we kept the solution weights with the minimum energy E_p over 20 such trials. The shape of the energy surface also creates difficulties if it has regions where the gradient in weight space is small. If the weights enter such a region, then $\partial E_p / \partial W_{lm}^{(k)} \approx 0$ for all $W_{lm}^{(k)}$, meaning that there is no way to distinguish the region from a true minimum. Whether or not the network weights escape such plateaus in the energy function depends strongly on the criterion that is used to stop the training. The two examples in Figures 2.6a and 2.7 are simple enough that 20,000 iterations in each of the 20 training trials was sufficient to produce the mappings illustrated in Figures 2.6d and 2.8.

One hidden layer unit was not sufficient to model the function $f(x)$ in Figure 2.6a. On the other hand, $f(x)$ was simple enough that two hidden layer units produced a good approximation of it; using ten hidden layer units in this case did not significantly improve the results. For the function in Figure 2.7, however, at least four units were necessary to 'capture' the primary structure in the data and increasing the number of hidden layer units to ten did improve the results. Too few units leads to a poor fit to the data and too many units does not improve performance while unnecessarily increasing the number of parameters in the network, perhaps leading to an over-fit of the data. If the underlying function of a problem is not known and there are only a limited number of samples from it, how does one choose the correct number of units in the hidden layers? Unfortunately, there is currently no clear-cut theoretical answer to this problem. In Section 2.4 we discuss one line of reasoning which suggests that enough hidden layer units should be implemented to solve the problem, i.e., produce a reasonable match to the data, and then techniques should be used to eliminate weights that do not appear to be important to the solution, thereby reducing the parameter set and the chances of an over-fit to the data.

For the network to best categorize novel data, i.e., generalize, the network must match the number of hidden layer units to the complexity of the problem as reflected in the training sample set. Baum and Haussler (1989) demonstrate that there is a relationship between the network size, the number of training samples and the ability of the resulting network to generalize:

> "Assume $0 < \varepsilon \leq 1/8$. We show that if $m \geq O((W/\varepsilon) \log (N/\varepsilon))$ random examples can be loaded on a feedforward network of linear threshold functions with N nodes and W weights, so that at least a fraction $1 - \varepsilon/2$ of the examples are correctly classified, then one has confidence approaching certainty that the network will correctly classify a fraction $1 - \varepsilon$ of future tests examples drawn from the same distribution. Conversely, for fully-connected feedforward nets with one hidden layer, any learning algorithm using fewer than $\Omega(W/\varepsilon)$ random training examples will, for some distributions of examples consistent with an appropriate weight choice, fail at least some fixed fraction of the time to find a weight choice that will correctly classify more than a $1 - \varepsilon$ fraction of the future test examples".

This statement drives home the point that if the training set size is too small relative to the number of weights, then there is no guarantee that the network will properly classify novel data.

The last point that we wish to make is that two hidden layers are sometimes better than one. In the mid-to-late 1980's there was some confusion about the mapping abilities of neural networks with one hidden layer (e.g., Lippman, 1987). The proofs of Cybenko (1989), Hornik et al. (1989), etc..., demonstrated that a neural network with one hidden layer can approximate any continuous function from R^n to R^m given enough units and the appropriate weights. However, a neural network with two hidden layers may solve some problems much more efficiently than a network with one hidden layer. For example, consider the function illustrated in Figure 2.7. We know that it cannot be reproduced exactly with one hidden layer of units. The best results that we obtained for a one hidden layer network in 20 trials with 20,000 iterations per trial using ten units in the hidden layer is illustrated in the last row of the right column of Figure 2.8. However, using the constructive scheme presented by Lippmann (1987), we designed the two hidden layer network in Figure 2.10 that reproduces the mapping exactly. There are 24 weights in this network, which is six less than the number of weights in the single hidden layer network with ten hidden units. Furthermore, in our first trial run using the backpropagation of error algorithm to train the network in Figure 2.10 initialized with random weights, we obtained the results in Figure 2.11. Clearly, for this mapping two hidden layers are better than one hidden layer.

2.3 Kohonen's Self-Organizing Feature Maps

The Kohonen self-organizing feature map (Kohonen, 1989) is a form of competitive learning with a network topology similar to Figure 2.3b, with two exceptions. First, the feature map does not explicitly use recurrent connections to decide the 'winner-take-all' output unit that remains active to the input. And second, the output units have a topology imposed on them, generally in two dimensions, that enables near-neighbor interactions to be defined (Figure 2.12).

In Grossberg's competitive learning scheme discussed in Section 2.2, the recurrent connections in the output layer provide a mechanism whereby the most active unit in the output shunts, i.e., turns off, the other units in the layer. If one is not concerned how the network is to find the output unit that responds most effectively to the input, then one simply has to calculate the output unit activities and then determine which one is the best according to some criterion (Kohonen, 1990; Rumelhart and Zipser, 1985; Intrator and Cooper, 1992). In Kohonen's self-organizational scheme, the activation function is

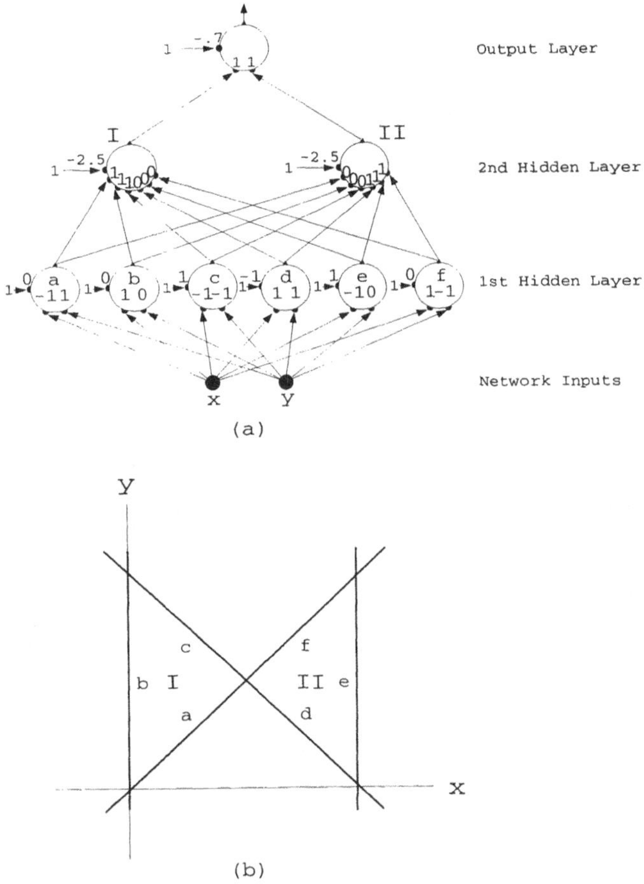

Figure 2.10: Using Lippmann's (1987) constructive technique, we explicitly build the multilayer perceptron with two hidden layers in (a) that reproduces the mapping $f(x, y)$ in Figure 2.7 exactly. (We used $u = u_d$ and $\sigma = \sigma_h$.) In (b) we illustrate the location of the hyperplanes for each first hidden layer unit in (a); the code letters are on the side of the hyperplane where the unit response is positive. Importantly, the two units in the second hidden layer respond only to inputs (x, y) that fall either within region I or II.

usually u_e, although u_d can be used, $W_{i0} = 0$ for all i, and the transfer function is $\sigma_l(a = 1, b = 0)$. Therefore, the output of the ith unit in the output layer is

$$o_i = \sigma_l(u_e(\mathbf{W}_i^{(1)}, 0, \mathbf{x})) = \sum_{j=1}^{n} (x_j^p - W_{ij}^{(1)})^2 \equiv E_i^p \qquad 38$$

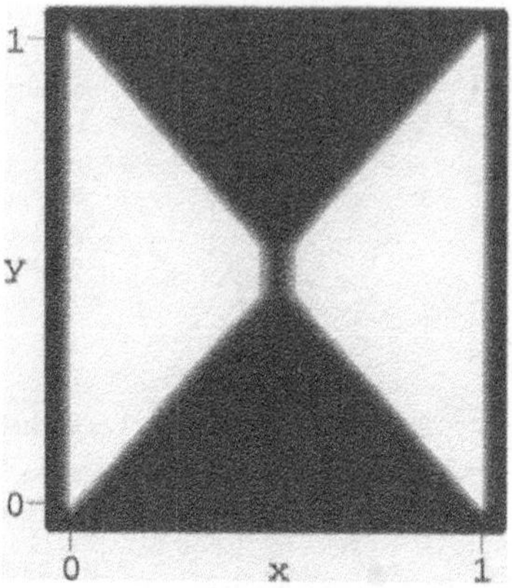

Figure 2.11: The two hidden layer multilayer perceptron approximation to $f(x, y)$ in Figure 2.7 after training the network with the backpropagion of error modification rule. The network architecture was identical to that of Figure 2.10. The error at the edge of the white area is due to use of $\sigma = \sigma_s$ during training and could be significantly reduced by letting $\alpha \rightarrow \infty$ in σ_s.

and the most effective response is the one with the smallest output, i.e., the unit with the weights $\mathbf{W}_i$ closest to $\mathbf{x}^p$ using Euclidean distance as the measure. Once the unit with the best response is determined according to some prescription, the weights $\mathbf{W}_i$ of the unit are updated so that they move closer to $\mathbf{x}_p$:

$$\Delta W_{ij}^{(1)} = \alpha \left(x_j^p - W_{ij}^{(1)} \right) \quad (j = 1, \ldots, n). \qquad 39$$

Patterns $\mathbf{x}^p$ are randomly input to the network and the weights are adapted according to (39). As the training cycle proceeds, the modification constant α is reduced to zero. Initially, α is large, allowing each unit's weights to make rapid adjustments to the input patterns. As each unit's weights $\mathbf{W}$ approach their respective input pattern $\mathbf{x}^p$, decrementing α allows the weights to move arbitrarily close to their target. Linde et al. (1980) and Gray (1984) show that the final locations of the weights $\mathbf{W}_i \in R^n$, and hence the densities of $\mathbf{W}^i$ in R^n, provide a measure of the probability distribution of the training patterns $\{\mathbf{x}^i \mid i = 1, \ldots, K\}$ in R^n. The network finds the structure in the data, i.e., the probability distribution of the training patterns in R^n, without supervision by minimizing an internal measure of network performance, i.e., the Euclidean distance between each input pattern and the closest set of unit weights $\mathbf{W}$.

In the above scheme each output unit ends up with its preferred input pattern, i.e., the one to which it is the closest. However, the output $o_i^{(1)}$ of unit i does not provide any information about the output $o_j^{(1)}$ of unit j $(i \neq j)$ because there is no interaction between

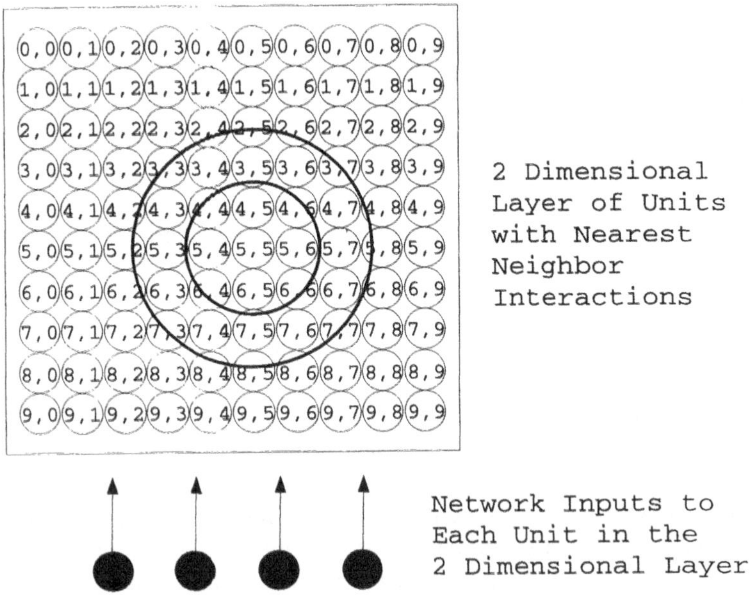

Figure 2.12: Kohonen's two-dimensional feature map network. The imposed topology allows the units to be labelled with their coordinates in, say, a Cartesian coordinate system. The Euclidean distance between the units can then be used as the metric to define the 'closeness' of two units. The concentric circles centered on a unit allow the near-neighbors of a unit to be defined at various times during training of the network.

different units in the output layer. Such is not the case, however, in the sensory areas of cortex in animals, where nearby neurons tend to respond to similar stimulii. Kohonen realized that to produce ordered maps, the network illustrated in Figure 2.3b must have a two-dimensional metric superimposed upon it and units must have strong near-neighbor interactions. Therefore, the network in Figure 2.3b is redrawn as a two-dimensional sheet (Figure 2.12) and the Euclidean distance d between two units, i.e.,

$$d(i,j) = \sqrt{(x_i - x_j)^2 + (y_i - y_j)^2} \qquad 40$$

where (x_i, y_i) and (x_j, y_j) are the coordinates of the two units, is used as the metric. Using such a topology, Kohonen changed the modification rule in (39) to include near-neighbor interactions:

$$\Delta W_{ij}^{(1)} = \begin{cases} \alpha\,(x^p - W_{ij}^{(1)}), & if \quad i \in N_i \\ 0, & if \quad i \notin N_i, \end{cases} \qquad 41$$

where N_i is a neighborhood of the unit i that is the closest to pattern $\mathbf{x}^p$. Therefore, all of the units sufficiently close to the unit being updated are updated as well. The result is a network that produces an ordered map of the weights $\mathbf{W}_i$ in the input pattern space such that the densities of the weights are a measure of the probability density of the patterns $\mathbf{x}^p$ in R^n and neighbors in the two-dimensional sheet of units are neighbors in the input pattern space.

During training of the self-organizing network, the neighborhoods N_i must initially cover a substantial fraction, e.g., greater than 0.5, of the output layer sheet; otherwise, the range of the interactions between output layer units may not be sufficiently broad to allow one map to organize across the entire output layer and several smaller disjoint maps may appear in the layer. Once the output layer units have organized on the large scale, the size of the neighborhoods N_i are gradually reduced so that only nearest-neighbor interactions remain; smaller neighborhoods enable local groups of units to respond to local structure in the input pattern probability distribution.

In Figure 2.13 we illustrate the evolution of the map that emerges in the 100 output layer units of Figure 2.12 when the probability distribution of the input patterns $\mathbf{x}^p \in R^2$ is given by Figure 2.7. The set of input patterns consists of the 200 patterns $\mathbf{x}^p$ such that $f(\mathbf{x}^p) = 1$ in (37). Therefore, the input patterns were uniformly distributed over the white areas in Figure 2.7. The initial values of the weights were random numbers between 0.0 and 0.5,

$$\alpha = \alpha_0 \left(1 - \frac{i}{N_{max}}\right), \quad \alpha_0 = 0.1, \qquad \text{42a}$$

and the neighborhoods N_i were circles with radii given by

$$r_N = r_0 \left(1 - \frac{i}{N_{max}}\right), \quad r_0 = 3.5 \qquad \text{42b}$$

(Kohonen, 1990). The input patterns were randomly selected and a total of N_{max}=200000 patterns were input to the network. The evolution of the resulting map is illustrated in Figure 2.13 from left-to-right and top-to-bottom at iteration 0, 100, 1000, 50000, 100000 and 200000. The final map is both organized, i.e., nearest-neighbor units are located in the same region of the input pattern space, and the distribution of the weights in R^2 provides a measure of the probability density function of the input patterns, i.e., uniformly distributed over the white areas in Figure 2.7.

2.4 Neural Networks and System Identification

Having concluded our introduction to the basic mechanics of neural network models applied in this monograph, we briefly return to the issue of the quality of solutions found by neural networks in general. We have briefly described the important issue of

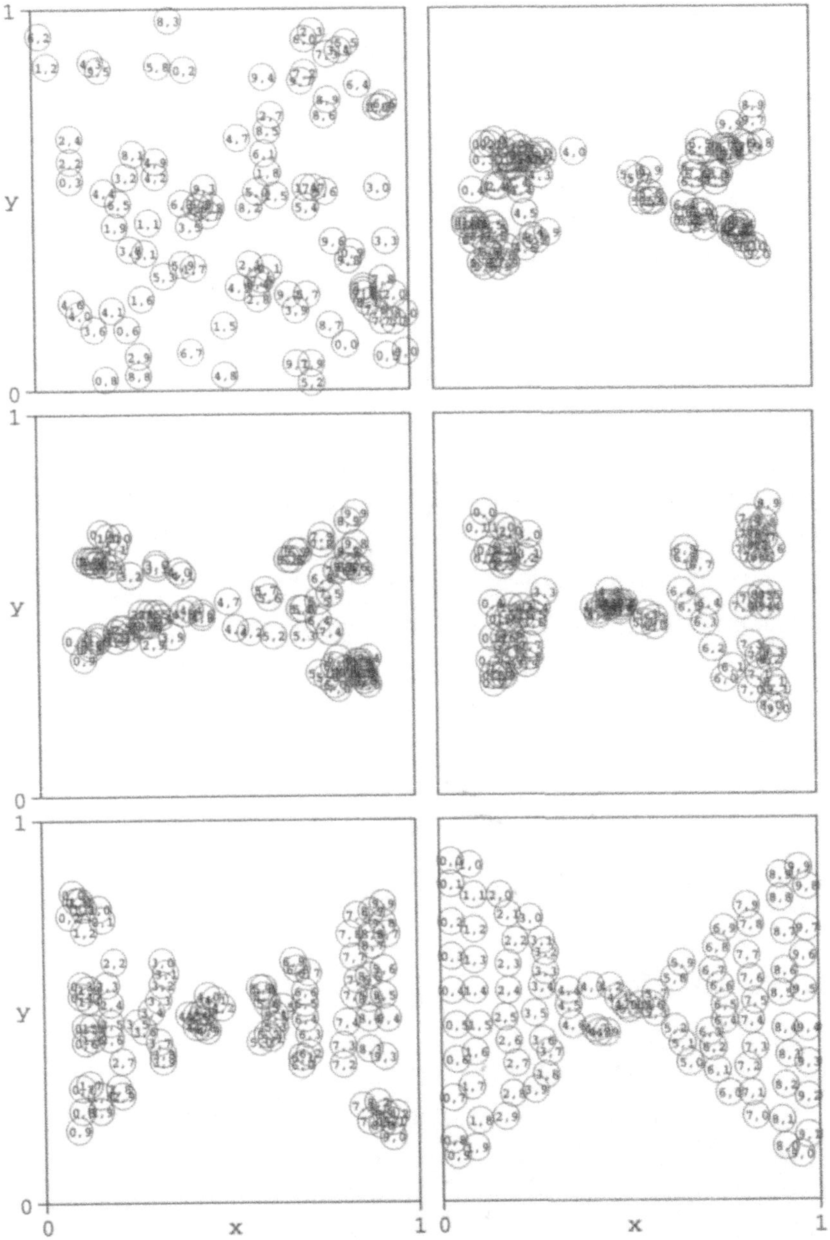

Figure 2.13: The evolution of the Kohonen feature map during training on samples taken from $f(x, y)$ in Figure 2.7; moving from left to right and top to bottom, the iteration times are 0, 100, 1000, 50000, 100000, and 200000. The location of each unit, labelled with its coordinates in the network topology given in Figure 2.12, is set by the unit's two weights.

cross-validation in applying neural networks to problems in pattern recognition. Recent papers have begun to treat these statistical issues (Solla, 1992) within a much more rigorous framework by bringing to bear techniques from the perspective of *System Identification* (Ljung and Sjoberg, 1992). From this statistical viewpoint, neural networks are a particular formalism for constructing models of mappings from a particular set of inputs to outputs. As such they are often described as 'non-parametric', despite the fact that they contain parameters for optimization. What is meant by this is that the order of the model is not so highly constrained as in more classical regression techniques, although models such as backward propagation may be thought of as a form of nonlinear regression. The thrust of the System Identification approach, as advocated by Ljung and Sjoberg, is to view neural networks as under-constrained, or ill-conditioned techniques, for solving problems. This means that in essence we allocate more resources for the problem than necessary, and allow the network to determine a model of sufficient order to solve the problem.

How then do neural networks avoid the pitfalls of overfitting data through too many parameters and obtaining a model worthy of Occam's Razor? As Ljung and Sjoberg point out, this is achieved in the neural network through regularization of the search process by applying one or many techniques, such as a cross-validation stopping criterion, a weight decay term in the modification rule, or a method for pruning or appending the number of nodes during training. To repeat their argument: to a first approximation, the variance in our estimate of a good model will be governed by the complexity of the model which we construct; the variance of our estimate for novel data will be described by:

$$E(\xi) = E(n^2)\,(1 + \frac{d}{N}), \qquad 43$$

where d is the number of parameters, N the number of data samples and $E(n^2)$ is the variance of the noise. When we regularize a cost function, such as the squared error over the network's responses, by adding a term such as weight decay, $\xi \rightarrow \xi + \alpha \sum (W - W_*)^2$, the variance is decreased according to:

$$E(\xi) = E(n^2)\left(1 + \frac{1}{N}\sum_i \frac{\sigma_i^2}{(\sigma_i^2 + \alpha^2)}\right), \qquad 44$$

where σ_i^2 are the eigenvalues of the Hessian matrix of the network response with respect to the parameters (weights **W**). Sjoberg and Ljung identify the term $d_\alpha = \sum_i \sigma_i^2 / (\sigma_i^2 + \alpha^2)$, which is at most equal to d (when $\alpha = 0$), as the effective number of parameters. This expression is not exact because we do not know the optimal W_* toward which to regularize, so in fact if α is too large there should be terms to describe this lack of knowledge and penalize the result. Nevertheless, this provides a conceptual framework for understanding regularization as a means of reducing the effective number of parameters in the network and preventing overfitting of the training data.

2.5 Areas of Current Research

We conclude this introductory chapter with a brief look at where the field of neural network research may be headed. A number of inter-related trends have been developing, which we group into three categories: (1) Ensemble Methods, (2) Chaos and Neural Networks, and (3) Biological Models. Let us define what we mean by each of these categories and show how these trends are inter-related.

2.5.1 Ensemble Methods

In the previous section, we described how the fields of Neural Network research and that of statistics are beginning to find common ground. Ensemble Methods represent yet another area in which this trend is being advanced. Simply defined, Ensemble Methods seek to improve the generalization accuracy of Neural Network classification, prediction or detection by combining the results of component networks trained to do essentially the same task. A number of papers have begun to treat the issue of neural networks from this perspective (Jacobs and Jordan, 1991; Perrone and Cooper, 1993a, 1993b) and have pointed out the connection between this approach and techniques used in the statistical literature. An important point made by Perrone is that this kind of technique is *not simply* an improved estimate of parameters (network connections) through multiple estimates as in regression, but rather an approach which generates an improved estimate in the mapping space itself, one closely related to Bootstrap and Jackknife techniques in statistics; the estimate in mapping or functional space, and *not* parameter space, is critical for neural networks because the large number of parameters in typical networks increases the likelihood of local minima and makes improved estimates through weighted averaging in parameter space less attractive. Indeed, even simple networks without adaptation can be used to achieve improved performance through ensemble estimates in mapping space, for example using sparse random receptive fields and a simple correlation mechanism for storage of training templates (Hansen et al., 1992).

2.5.2 Biological Realism and Chaos

In some sense the field of neural network research is at a crossroads. Artificial neural network models of the perceptron type, i.e., models incorporating a weighted sum of inputs with a nonlinear sigmoidal transfer function, are now well understood, and it seems that in order to reach the next generation of algorithms, researchers are beginning to look more closely at the underlying biology to develop more powerful computational units and architectures. Indeed, sophisticated biophysical models of synaptic plasticity are beginning to emerge (De Schutter and Bower, 1993), which look at the dynamics of ion channels and receptors. Models such as these are useful for studying effects such as long-term potentiation (LTP), which is believed to be a physiological manifestation of learning. At the same time, one of the themes to emerge from biology is the chaotic nature of the nervous system (Freeman, 1988; Lo and Principe, 1989), and many researchers are now beginning to look at the brain as a highly complex dynamical system in which individual cells and groups of cells exhibit deterministic chaos.

These two areas of research, chaos and biological realism, are beginning to move toward each other in two distinct ways. As a signal processing tool, the study of chaotic attractors has proven to be a powerful method for extracting information from noisy and apparently random signals, so that on the one hand, some researchers have already begun to use chaotic signal processing techniques, such as phase-space reconstruction, as a preprocessing step for standard artificial neural network models (Tumey et al., 1991). At the same time working in the other direction some researchers have already begun to incorporate more sophisticated biological knowledge into their models (Chay, 1991; Aihara, 1991; Adachi et al., 1991; Freeman, 1989). These detailed models often lead to the kind of chaotic behavior found in the nervous system. Researchers, such as Freeman, have realized that if the brain can represent information as a chaotic attractor, then artificial models for pattern recognition should also exploit this. Indeed, this direction offers new hope for an entirely different representational paradigm and, perhaps, for a generation of more powerful processing algorithms.

Acknowledgements

E. E. Clothiaux is supported by an appointment to the Global Change Distinguished Postdoctoral Fellowships sponsored by the U.S. Department of Energy, Office of Health and Environmental Research, and administered by the Oak Ridge Institute for Science and Education.

C. M. Bachmann is supported by a grant through Dr. Thomas McKenna at the Office of Naval research (N0001493AF00002) and core funds at the Naval Research Laboratory (53-1501-3).

References

Adachi, M., Aihara, K., Kotani, M. 1991 "Pattern Dynamics of Chaotic Neural Networks with Nearest-Neighbor Couplings", IEEE International Symposium on Circuits and Systems, 2, p. 1180-1183.

Aihara, K. 1991 "Chaotic Dynamics in Nerve Membranes and its Modelling with an Artificial Neuron", IEEE International Symposium on Circuits and Sytems, 3 p. 1457-1460.

Amari, S.-I. 1977 "Neural Theory of Association and Concept-Formation". Biological Cybernetics, 26 p. 175-185.

Anderson, J.A. 1972 "A Simple Neural Network Generating an Interactive Memory", Mathematical Biosciences, 14 p. 197-220.

Anderson, J.A., Silverstein, J.W., Ritz, S.A., Jones, R.S. 1977 "Distinctive Features, Categorical Perception, and Probability Learning: Some Applications of a Neural Model", Psychological Review, 84 p. 413-451.

Anderson, J.A., Rosenfeld, E. 1988 "Neurocomputing: Foundations of Research", MIT Press, Cambridge, MA.

Baum, E.B., Haussler, D. 1989 "What Size Net Gives Valid Generalization?", Neural Computation, 1 p. 151-160. Block, H.D. 1962 "The Perceptron: A Model for Brain Functioning", Reviews of Modern Physics, 34, p. 123-135.

Carpenter, G.A., Grossberg, S. 1987 "A Massively Parallel Architecture for a Self-Organizing Neural Pattern Recognition Machine", Computer, Vision, Graphics, and Image Processing, 37 p. 54-115.

Carroll, S.M., Dickinson, B.W. 1989 "Construction of Neural Nets Using the Radon Transform", IEEE International Joint Conference on Neural Networks, Washington, DC, 1 p. 607-611.

Chay, T.R. 1991 "Complex Oscillations and Chaos in a Simple Neuron Model", IEEE International Joint Conference on Neural Networks, Seattle, WA, 2, p. 657-662.

Cooper, L.N. 1973 "A Possible Organization of Animal Memory and Learning", In Proceedings of the Nobel Symposium on Collective Properties of Physical Systems, B. Lundquist and S. Lundquist (Eds.), Academic Press, New York, p. 252-264.

Cybenko, G. 1988 "Continuous Valued Neural Networks with Two Hidden Layers are Sufficient", Technical Report, Dept. of Computer Sciences, Tufts University, Medford, MA.

Cybenko, G. 1989 "Approximation by Superpositions of a Sigmoidal Function", Mathematics of Control, Signals, and Systems, 2(4), p. 303-314.

DARPA 1988 "DARPA Neural Network Study", AFCEA International Press, Fairfax, VA.

Dayhoff, J.D. 1990 "Neural Network Architectures: An Introduction", Van Nostrand Reinhold, New York.

De Schutter, E., Bower, J.M. 1993 "Sensitivity of Synaptic Plasticity to the Permeability of NMDA Channels: A Model of Long-Term Potentiation in Hippocampal Neurons", Neural Computation, 5, p. 681-694.

Freeman, W.J. 1988 "Strange Attractors that Govern Mammalian Brain Dynamics Shown by Trajectories of Electroencephalographic (EEG) Potential", IEEE Transactions on Circuits and Sytems, 35:(7), p. 781-783.

Funahashi, K.-I. 1989 "On the Approximate Realization of Continuous Mappings by Neural Networks", Neural Networks, 2, p. 183-192.

Gibson, G.J., Cowan, C.F.N. 1990 "On the Decision Regions of Multilayer Perceptrons", Proceedings of the IEEE, 78:(10) p. 1590-1594.

Giles, C.L., Sun, G.Z,, Chen, H.H., Lee, Y.C., Chen, D. 1990 "Higher Order Recurrent Networks and Grammatical Inference", In Advances in Neural Information Processing Systems, 2, D. S. Touretzky (Ed.), Morgan Kaufmann, San Mateo, CA.

Giles, C.L., Miller, C.B., Chen, D., Sun, G.Z., Chen, H.H., Lee, Y.C. 1992 "Extracting and Learning an Unknown Grammar with Recurrent Neural Networks", In Advances in Neural Information Processing Systems, 4, J.E. Moody, S.J. Hanson, and R.P. Lippmann (Eds.), Morgan Kaufman, San Mateo, CA.

Girosi, F., Poggio, T. 1989 "Representation Properties of Networks: Kolmogorov's theorem is irrelevant", Neural Computation, 1:(4) p. 465-469.

Gray, R.M. 1984 "Vector Quantization", IEEE ASSP Magazine, 1 p. 4-29.

Grossberg, S. 1973 "Contour Enhancement, Short Term Memory, and Constancies in Reverberating Neural Networks", Studies in Applied Mathematics, 52 p. 213-257.

Grossberg, S. 1976 "Adaptive Pattern Classification and Universal Recoding: I. Parallel Development and Coding of Neural Feature Detectors", Biological Cybernetics, 23, p. 121-134.

Hansen, L.K., Liisberg, C., Salamon, P. 1992 "Ensemble Methods for Handwritten Digit Recognition", In Neural Networks for Signal Processing II - Proceedings of the 1992 IEEE Workshop, Copenhagen, Denmark, p. 333-342.

Hebb, D.O. 1949 "The Organization of Behavior", Wiley, New York. Hecht-Nielsen, R. 1990 "Neurocomputing", Addison-Wesley, Reading, MA.

Hopfield, J.J. 1982 "Neural Networks and Physical Systems with Emergent Collective Computational Abilities", Proceedings of the National Academy of Sciences, 79, p. 2554-2558.

Hornik, K., Stinchcombe, M., White, H. 1989 "Multilayer Feedforward Networks are Universal Approximators", Neural Networks, 2, p. 359-366.

Hush, D.R., Horne, B.G. 1993 "Progress in Supervised Neural Networks: What's New Since Lippmann", IEEE Signal Processing Magazine, 10:(1) p. 8-39.

Intrator, N., Cooper, L.N. 1992 "Objective Function Formulation of the BCM Theory of Visual Cortical Plasticity: Statistical Connections, Stability Conditions", Neural Networks, 5, p. 3-17.

Jacobs, R.A., Jordan, M.I. 1991 "Adaptive Mixtures of Local Experts", Neural Computation, 3, p. 79-87.

Jones, L.K. 1990, "Constructive Approximations for Neural Networks by Sigmoidal Functions", Proceedings of the IEEE, 78:(10) p. 1586-1589.

Jones, L.K. 1992 "A Simple Lemma on Greedy Approximation in Hilbert Space and Convergence Rates for Projection Pursuit Regression and Neural Network Training", The Annals of Statistics, 20:(1) p. 608-613.

Knight, K. 1990 "Connectionist Ideas and Algorithms", Communications of the ACM, 33:(11). p. 59-74.

Kolmogorov, A.N. 1957 "On the Representation of Continuous Functions of Several Variables by Superposition of Continuous Functions of One Variable and Addition", Dokl. Akad. Nauk SSSR, 114, p. 953-956.

Kohonen, T. 1989 "Self-Organization and Associative Memory", Springer-Verlag, Berlin, Germany.

Kohonen, T. 1990 "The Self-Organizing Map", Proceedings of the IEEE, 78:(9), p. 1464-1480.

Kosko, B. (Ed.) 1992 "Neural Networks for Signal Processing", Prentice Hall, Englewood Cliffs, NJ. Lau, C. (Ed.) 1992 "Neural Networks: Theoretical Foundations and Analysis. IEEE Press, New York.

Linde, Y., Buzo, A., Gray, R.M. 1980 "An Algorithm for Vector Quantization", IEEE Transactions on Communications, COM-28, p. 84-95.

Lippmann, R.P. 1987 "An Introduction to Computing with Neural Nets", IEEE ASSP Magazine, 4, p. 4-22.

Lorentz, G.G. 1976 "The 13th Problem of Hilbert", In Mathematical Developments Arising from Hilbert Problems", F. E. Browder (Ed.), American Mathematical Society, Providence, RI, p. 419-430.

Ljung, L., Sjoberg, J. 1992 "A System Identification Perspective on Neural Nets", Neural Networks for Signal Processing II - Proceedings of the 1992 IEEE Workshop, Helsingor, Denmark, p. 423-435.

McClelland, J.L., Rumelhart, D.E., and the PDP Research Group 1986 "Parallel Distributed Processing: Explorations in the Microstructure of Cognition, Volume 2: Psychological and Biological Models, MIT Press, Cambridge, MA.

Moore, B., Poggio, T. 1988 "Representations Properties of Multilayer Feedforward Networks", In Abstracts of the First Annual INNS Meeting, Pergamon Press, New York, p. 502.

Parlos, A., Atiya, A., Chong, K., Tsai, W., Fernandez, B. 1991 "Recurrent Multilayer Perceptron for Nonlinear System Identification", IEEE International Joint Conference on Neural Networks, Seattle, WA, 2, p. 537-540.

Perrone, M.P., Cooper, L.N. 1993 "Learning from What's Been Learned: Supervised Learning in Multi-Neural Network Systems", Proceedings of the World Congress on Neural Networks, Portland, OR, p. 354-357.

Perrone, M.P., Cooper, L.N. in press "When Networks Disagree: Ensemble Methods for Neural Networks", In Neural Networks for Speech and Image Processing, R. J. Mammone (Ed.), Chapman Hall.

Pollack, J.B., 1991 "The Induction of Dynamical Recognizers", Machine Learning, 7, p. 227-252. Rosenblatt, F. 1958 "The Perceptron: A Probabilistic Model for Information Storage and Organization in the Brain", Psychological Review, 65, p. 386-408.

Rumelhart, D.E., Hinton, G.E., Williams, R.J. 1986a "Learning Internal Representations by Error Propagation", In Parallel Distributed Processing: Explorations in the Microstructures of Cognition, Volume 1: Foundations, D.E. Rumelhart and J.L. McClelland (Eds.), MIT Press, Cambridge, MA, p. 318-362.

Rumelhart, D.E., Hinton, G.E., Williams, R.J. 1986b "Learning Representations by Back-Propagating Errors", Nature, 323 p. 533-536.

Rumelhart, D.E., McClelland, J.L., and the PDP Research Group 1986c "Parallel Distributed Processing: Explorations in the Microstructure of Cognition", Volume 1: Foundations, MIT Press, Cambridge, MA.

Rumelhart, D.E., Zipser, D. 1985 "Feature Discovery by Competitive Learning", Cognitive Science, 9, p. 75-112.

Solla, S.A. 1992 "Capacity Control in Classifiers for Pattern Recognition", In Neural Networks for Signal Processing II- Proceedings of the 1992 IEEE Workshop, Helsingor, Denmark, p. 255-266.

Stinchcombe, M., White, H. 1989 "Universal Approximation Using Feedforward Networks with Non-Sigmoid Hidden Layer Activation Functions", IEEE International Joint Conference on Neural Networks, Washington, DC, 1, p. 613-617.

Tumey, D.M., Morton, P.E., Ingle, D.F., Downey, C.W., Schnurer, J.H. 1991 "Neural Network Classification of EEG Using Chaotic Preprocessing and Phase Space Reconstruction", Proceedings of the 1991 IEEE Seventeenth Annual Northeast Bioengineering Conference, p. 51-52.

Widrow, B., Hoff, M.E. 1960 "Adaptive Switching Circuits", IRE WESCON Convention Record, 4, p. 96-104.

Williams, R.J., Zipser, D. 1988 "A Learning Algorithm for Continually Running Fully Recurrent Neural Networks", ICS Report 8805, University of California, San Diego, CA.

Yao, Y., Freeman, W.J. 1989 "Pattern Recognition in Olfactory Systems: Modeling and Simulation", IEEE International Joint Conference on Neural Networks, Washington, DC, 1, p. 699-704.

Eugene E. Clothiaux, The Pennsylvania State University, 503 Walker Building, University Park, PA 16802, USA.

Charles. M. Bachmann, Airborne Radar Branch, Radar Division, Naval Research Laboratory, Washington, DC, USA.

Chapter Three

NEUROCLASSIFICATION OF SPATIAL DATA

Stan Openshaw

3.0 Towards a Computational Geography

Without any doubt geography is now in the midst of its third quantitative revolution (cf the statistical revolution in the early 1960s, the mathematical modelling revolution in the early 1970s, and now the neurocomputing revolution in the early 1990s). In common with many other areas of science there is a rapidly growing interest in the application of neurocomputing methods. In many ways the driving force is external to the subject in that the tools are being imported rather than developed indigenously. The new tools are also replacements or complements for, or to, existing methods. The general justification is the promise of an improvement in performance and efficiency, fewer critical assumptions, greater ease in handling hard problems, an expansion of the applicability of quantitative and computer based tools and eventually, automation. It is noted that neurocomputing is just one source of new quantitative tools for geography in the 1990s (Openshaw, 1992a).

It is quite clear that many of the quantitative analysis and modelling problems in geography class as 'hard' problems. For example, the seemingly simple task of building a spatial regression model for use in a GIS has not yet been satisfactorily solved. The problems of spatial dependency, modifiable areal units, and spatial heterogeneity remain as important barriers to progress, and they may never adequately be solved using conventional statistical methods. Similarly, the available 1970s mathematical model building technology (ie Wilson, 1974) cannot cope with Weaver's type III system; systems of organized complexity. The world of chaos seemingly confounds the statistical mechanics based tools of the early 20th century physicist when applied to realistically complex geographical systems. We can, it seems, no longer find, or be certain, that deterministic systems of equations based on the best available geographical theories are adequate for what is expected from them. Neuromodelling provides in both cases the basis for a radically different and new approach to solving existing, old, problems and for tackling new, hitherto, insolvable ones. It comes with the unique promise that it will probably work, at a cost. For example, Openshaw (1993) demonstrates some neurocomputing based spatial interaction models that performed

B. C. Hewitson and R. G. Crane (eds.), Neural Nets: Applications in Geography, 53–70.

better than conventional entropy- maximizing models - albeit requiring three orders of magnitude increase in CPU times.

The secret, then, if there really is one, is to seek to replace a statistical and analytical mathematics based quantitative geographic technology by a computational one, based initially perhaps on neurocomputing methods but broadening to embrace all of the applicable Artificial Intelligence (AI) tools. The 1990s will witness an historically unprecedented speed-up in computer hardware. Within a decade virtually any problem that can be solved by computation will be solvable and practical to solve, provided the computation task is a highly parallel one. In essence neurocomputing is computationally parallel, inspired as it is by the functioning of the massively parallel human brain. Geography stands, today at the beginning of its last great revolution, that of computational geography. The hope is expressed that Geography will, over the course of the next 50 years, link all its subdisciplines together in a process model based cybernetic framework capable of providing a reasonable and scientific based understanding of geographical systems (Openshaw, 1991, 1992b).

The specific interest here within computational geography is that of applying simple neurocomputing methods to discover structure and pattern via the exploratory analysis of spatial data. The objective is that of spatial data description or regional description, one of the most basic and under-rated of all geographical analysis tasks. The methods of interest are those of unsupervised neural networks sometimes confusingly called competitive learning.

The task of providing simplified descriptive summaries of seemingly massively complex multivariate spatial databases was one faced by the early quantitative geographers in the 1960s. The GIS revolution of the mid-1980s and the subsequent explosion in spatial information (it is currently believed that the number of computer databases is doubling every 5 years) emphasizes the importance of this descriptive task. The aims are to simplify, describe, reduce, and add-value to spatial data by identifying (or discovering or uncovering) the major features contained in the data. Previously, geographers used numerical taxonometric methods (Sokal and Sneath 1963; Spence and Taylor, 1970; Everitt, 1974; Openshaw, 1976). Whilst basic classification methods are now widely available in popular statistical packages, their capabilities are today not greatly different from 30 years ago. They are, sadly, not really adequate for making the most of the new data rich environments of the 1990s and beyond, and are thus inadequate for extracting maximum geographic value from the many new data sources that are appearing in many countries as a result of the computerization of all routine management functions.

The research challenge now is to do something useful with all this data at a time when data supply in many areas has often far outstripped either a priori theories to test or even any relevant prior knowledge. Quite often the data explosion has occurred in areas that previously attracted little academic interest, but in which, due to changing circumstances, there is often an imperative for analysis; for example, real-time crime data or disease databases.

In this increasingly data abundant era, tools are needed that can help people cope. Data reduction via simple descriptive summaries or characteristic profiles, or idealized typologies are suddenly greatly in demand as researchers and managers ask fairly simple but also very vague questions, such as "what patterns exist" or "where are the anomalies" or "in a nut shell what on earth is happening". People cannot 'see' multivariate patterns in massive spatial databases. Without new tools they will find it increasingly hard to cope with the new cyberspaces within which they increasingly have to operate. Clearly then classification methods are a class of useful data coping and visualizing tools that are able to answer at least some of these broad questions. In some areas, such as census analysis, geodemographic classifications attempt to provide this over-view and have been very successful as an applied tool. The question now, is can we not do considerably better?

Openshaw (1989, 1993b) argues that most of the available census classification tools represent 1960s technology clothed in 1990s computer software. All that has changed is the ease of use, access to the technology, and size of the largest application. At the same time, it is now possible to do much better, at least in theory. Computers are about 10,000 times faster than 20 years ago and are set to become about 10,000 times faster again by the late 1990s. Sustained terraflop computer speeds on massively parallel processors are confidently expected by the late 1990s. At the same time most data sets for most countries are reaching asymptotic limits in their maximum size; for example, in Britain the largest ever possible people databases is about 53 million. In the late 1960s, geographers had large data set size values of 2 digits, by the late 1970s it had reached 130,000, today it has asymptoted at 53 million and cannot get much bigger. The challenge is no longer just to survive with multi-megabyte databases by applying the simplest of methods at the cost of quality, but of devising the best possible methods for use with gigabyte databases on terraflop hardware. There is no doubt that this will happen, although when is still uncertain. Equally there is no real doubt that new tools will be needed and that they will be based on AI and neurocomputing but again what precisely are the best ones is still an open question.

The basic rule of computational geography has already been outlined. In essence anything that can be solved by computation is or will become a practicable proposition, no matter how infeasible it appeared five years ago, or even now. But you can solve problems by throwing computing power at them only if the algorithms are simple enough, perhaps dumb enough, to be explicitly suitable for parallel computation. In short we can trade compute power for efficiency and flexibility only if we seek to become clever by first becoming dumber and less sophisticated in what we seek to do. Neurocomputing is a classic example of this philosophy although many of its advocates might well not consider what they propose as being dumber than conventional practice. In this context, you become dumber by developing often seemingly 'simple' methods that can exploit parallel computing and, of course, by abandoning conventional technology that may well have a scientific pedigree of 100 years or more. In the short-term considerable faith is needed, in the longer term success is probably assured if you can develop the 'right' tools that can best exploit the new data and new computing environments; the two are closely related although this is often forgotten. Currently, the available methods assume implicitly slow expensive computing and small data set sizes; neither assumptions are true any longer.

3.1 Whither Neuroclassification?

The fundamental attraction of what is termed here neuroclassification is that it seeks to offer a new approach to the classification of explicitly spatial data based on developments in neurocomputing. Notwithstanding all that has been written so far, the first question that should be asked is "why?" Unless you can convincingly answer this question in an honest fashion, then you should not proceed to experiment with any of the methods described here. Indeed you should seek counselling because you are probably a victim of the highly infectious neural net virus. It is a very dangerous condition because of the trance like tunnel vision it can engender in susceptibles. Possible answers to the "why" question are as follows: (1) its fashionable (2) its different (3) its flexible (4) its better (5) its automatic, and (6) its smart technology.

Lets be quite clear here. There is no real point in struggling to master possibly unfamiliar technology clothed in strange jargon with a mix of the electrical engineer, the neurocognitive scientist, and the biologist; just because it is fashionable to work in the neural net area. Personal curiosity and research training are worthy objectives. However there are limits to both, and the latter tends almost by definition to be restricted to established technology. One of the key characteristics of neuroclassification is that it is both not yet established and still experimental in an empirical sense. Indeed this is one of its greatest attractions but this is really only a temporary situation. Just because neuroclassification offers a different route to classifying spatial data is not by itself a strong justification for applying it. Flexibility is important but is not sufficient by itself. The only real justification is the belief, or feeling, that it might well yield better classifications that are 'smarter' by being more context sensitive to the nature of the data and the peculiarities of its application.

Automation is clearly important but only in the longer term. De-skilling is likewise important but mainly to cope with the prevailing skill shortage (how many good, empirically experienced, spatial taxonomists are there in the world right now?), the explosion in the numbers of spatial data sets to analyze, and the need to speed-up the analysis process so that real time operations are possible. Likewise being 'smart' (cf.

being dumb) is clearly an attractive attribute to possess if it can be proven, but is it sufficient by itself? How 'smart' do you have to be before it really becomes useful? Indeed 'slightly smart' systems are usually a pain. Being better is much more useful, but how do you prove it? Also, what grounds are there for even believing that neuroclassifiers might be even slightly better let alone sufficiently significantly better to justify the effort required in applying them.

To be fair, it is also early days as yet for neuroclassifiers and a mix of realism, enthusiasm, faith, and luck are very vital ingredients. It is with these thoughts in mind and against a more general background of computational geography, that this chapter attempts to capture and convey the excitement and spirit of adventure and enthusiasm that today surrounds neurocomputing and focus it on the neuroclassification theme. Section 3 briefly reviews different neuroclassifier architectures before concentrating on one of these, the Kohonen self-organizing map. Section 4 provides a detailed algorithmic description with a source listing provided in Appendix II. Section 5 provides a simple empirical application and section 6 a discussion of how the technology might be developed further.

3.2 Review of Potential Neuroclassifier Architectures

The single most abiding feature shared by all neuroclassifiers is their inherent simplicity. Stripped of brain biology jargon the technology is wonderfully straightforward. Indeed it has to be, because it is widely recognized that the complexity of the brain is due to the density and interconnectedness of the vast numbers of neurons, and is not due to the complexity of the individual neurons themselves.

In a general way classification of objects is one of the most fundamental tasks for the human brain. We all possess an amazing ability to recognize objects, to extract features of interest from a highly complicated scene and to summarize high dimensional complex sensory data simply and abstractly. It is not surprising, therefore, to find many references to classification and neurobiological terms that also imply classification or something closely related to it in the literature on artificial neural nets. Simpson (1990) identifies 17 different unsupervised learning paradigms; he calls them paradigms because there are multiple variants of each of them. Unsupervised learning is a process of self-organization whereby the neural net discovers the natural structure of data by itself, using only samples of data and internal control mechanisms. The net trains itself to learn or uncover the principal features present in the data presented to it. This is clearly a classification function but how many numerical taxonomists would have considered what they have traditionally done as an unsupervised training process; or that the classification of a new case is a form of associative recall operation from a memory; or that the classification task is simultaneously also a data compression, data mapping, and data encoding function. The neurocomputing literature contains many interesting references to procedures and processes which perform what might be regarded as a classification function. It follows, therefore, that the review proposed here is both partial and incomplete.

A recurrent key theme in unsupervised nets is that of self-organization. If the net is to identify common features in its inputs, it must somehow be able to determine what these features are for itself. The concept of self-organization is biologically motivated, in that there is evidence that parts of the brain seem to operate in that way. Perhaps the principal mechanism whereby the brain self-organizes is via competitive and cooperative processes. These are not new ideas. Competitive learning was introduced by Grossberg (1972) and Malsburg (1973), amongst others and has been developed in various ways. Attention here is focused on four particular types of unsupervised learning architecture that may be of greatest relevance to neurospatial classification; the reader interested in a broader typology is referred to Simpson (1990) Chapter 5.

3.3 Competitive Learning Nets

The simplest architecture is the competitive net described so well by Rumelhart and Zipser (1985). However, there is a long ancestry including Rosenblatt (1962), Grossberg (1976), and Kohonen (1984). In its simplest form a set of processing elements fight it out to represent data using a measure of dissimilarity. The single winner updates its weights. Over a period of time the weights attached to each processing element will provide a summary of any structure or regularities existing in the data on which it was trained. This is clearly very similar to cluster analysis and does little more than identify similarity classes. However, the user has much more control over the training process and can deliberately interfere with the equiprobable distribution of weights, in order to handle better the nature of spatial data. How this is achieved is discussed later.

Rumelhart and Zipster (1985) have extended this basic model to include multiple layers of processing elements, and within each layer there can be different groups of processing elements (exclusive or non exclusive). The hope is that the different layers will identify different forms of structure, applying varying degrees of generalization and simplification. This is somewhat fanciful and network design is highly subjective but the underlying objectives are seemingly, extremely, worthwhile.

3.4 Self-Organizing Map

A second major type of architecture is that based on the self-organizing map (SOM) of Kohonen (1984). This is basically a multiple winner competitive net. The processing elements have a spatial structure to them, typically they are arranged on a two-dimensional grid. Once a winner is found then updating takes place of the winning processing element and those within a certain critical neighborhood of it, whilst those further away are inhibited. This might also be termed a competitive-cooperative architecture with lateral inhibition. This critical neighborhood is gradually reduced during training. Once convergence is complete, the weights associated with each processing element define a Voronoi tessellation of the data space. This SOM provides in effect a non-parametric pattern classification. It is widely regarded as one of the most useful of unsupervised neural nets and its use as a neuroclassifier is the subject of much of the remainder of the chapter.

3.5 Adaptive Resonance Theory

A third category of potentially useful neuroclassifier is the Adaptive Resonance Theory (ART 1,2) variants developed by Carpenter and Grossberg (1987a, b). They are essentially nearest neighbor classifiers but modelled on the functioning of the brain. The patterns they store are regarded as exemplars and when a 'new' pattern is found, then a new exemplar (ie group) is created for it. These methods are essentially analog pattern classifiers but there is no reason at all why they would not also perform well with spatial data. It can learn fast (suitable for on-line use) or slowly and can dynamically adapt to the structure found in the data via a vigilance process.

A subsequent variant ARTMAP (Carpenter and Grossberg, 1991) seems to be an even better prospect for further development. The principal problem for the geographer (and many others) is simply that of understanding the neurobiological technical language sufficiently well to understand precisely what is going on. Maybe an example will illustrate this point. Carpenter and Grossberg (1987) write "when a mismatch attenuates STM activity across F1, the total size of the inhibitory signal from F1 to A is also attenuated. If the attenuation is sufficiently great, inhibition from F1 to A can no longer prevent the arousal source A from firing."

3.6 Associative Memory Nets

A fourth category of unsupervised nets are mainly associative memory devices. That is they store patterns in such a way that they can be retrieved using patterns or parts of patterns or noisy parts of patterns as retrieval keys; examples are Hopfield nets (Hopfield, 1982, 1984), Bidirectional Associative Memory (BAM) (Kosko, 1988), Temporal Associative Memory (TAM) Amari, 1972), Linear Associative Memory (LAM) (Kohonen, 1977) and Sparse Distributed Memory (SDM) Kanerva, 1988). These methods provide a different route to spatial classification, which, if it could be operationalized, would involve building libraries of spatial pattern types and their extensions. It might well become possible one day to think in this way, i.e. of encoding concepts as spatial pattern exemplars (and vice versa) but sadly not at present. These sophisticated pattern matching methods are likely to become much more important later on than at present.

3.7 Comparisons With Conventional Classifiers

Finally it is noted that there are some strong similarities between neuroclassifiers and conventional taxonomic methods. Any superficial comparison would immediately suggest that the neurons in a competitive net are fighting to represent data cases in a manner similar to the cluster centroids in the conventional iterative relocation procedure; indeed, they may share the same dissimilarity criteria. Again it can be argued that a K-means algorithm is similar to Grossberg's ART-2 system. Doubtlessly several other superficial similarities might be found. However, there is a very important point here. In nearly all cases neuroclassifiers are functionally quite different, they are not computationally identical; indeed far from it.

The net learning mechanisms have no conventional taxonomic equivalent (although one could be manufactured). The lateral inhibition characteristics of a Kohonen self-organizing map based on a spatial pattern of cooperating and non-cooperating neurons has no conventional taxonomic equivalent (and one could not be manufactured). The basic learning process is inherently different (although conventional methods might well benefit here) and is much more likely to avoid suboptima than a conventional single more heuristic process. Additionally, neuroclassifiers escape many of the problems that afflict conventional methods. For instance, a standard single move heuristic classifier based on principal component scores has the following problems: its old technology and might get stuck in a local suboptimum, the data are assumed to be normally distributed and free from outliers (e.g. census data is usually J shaped), relationships that matter are linear (else the principal component transformation filters them out), each case has the same weight, the data has a uniform level of accuracy and reliability, spatial autocorrelation does not exist (else the principal component scores will be biased), and that cluster morphology is multivariate spherical and minimum variance in an euclidean distance norm. None of this need be applicable to a neuroclassifer applied appropriately.

Despite problems, it is clear that conventional classifiers do work. Maybe the users compensate for the failures inherent in the technology; maybe success is more apparent than real because of an absence of benchmarked results. The view here is that neuroclassifiers may well help if used in an appropriate fashion within the context of good research practice.

3.8 Kohonen's Self-Organizing Map

Kohonen's self-organizing map is one of the most interesting of all the competitive neural nets. Its fascination results from the realization that self-organization is a very powerful neural process and that parts of the brain certainly seem to operate in a similar fashion. To the geographer the main attraction is initially spurious; i.e. the word 'map', but subsequently the spatial concepts involved in self-organizing maps are extremely interesting for a variety of applications. As a neurospatial classifier, the self-organizing map offers a number of attractions: (1) simplicity in algorithmic design; (2) ability to handle immense complexity; (3) nice mathematical properties; (4) user induced flexibility; and (5) a plausible degree of biological inspiration. All are important. Compared with conventional classifiers, it is very simple. At the same time, its self-organizing nature offers immense benefits of local adaptiveness and emergent behavior. The nice mathematical properties mainly relate to the topological preserving nature of the results. Finally, user induced flexibility reflects the lack of any statistical over-sophistication that might induce delusions of grandeur. The user is free to experiment and include whatever is considered important without too many constraints on avenues of possible actions, although the onus is firmly with the researcher to demonstrate the validity of the results (Openshaw, 1992b).

3.8.1 A Basic Algorithm

The basic algorithm is really very simple (Beale and Jackson (1990), Caudill and Butler (1992) provide introductory descriptions). However it is important to remember that there are many possible variations and that in any particular application some experimentation will be worthwhile.

A basic Kohonen network algorithm can be described as follows:

Step 1. Initialization. Define geometry, dimensionality, and size of neuron array.
Step 2. Each neuron has a vector of M weights. Set these weights to some initial value, usually random values.
Step 3. Select a data case that also has variable values and apply any relevant measurement noise to the data.
Step 4. Find whichever neurons is 'nearest' to the data case under consideration.
Step 5. 'Update' the vectors of M weights for all the neurons in the topological neighborhood of the winning neuron, otherwise leave alone.
Step 6. Reduce learning parameter and neighborhood weights by a very, very small amount.
Step 7. Repeat steps 3 to step 6 until convergence, typically a large number of times; viz 100,000 or a million or more.

This basic algorithm is a computationally simple version that generally might be expected to produce very reliable maps. It is easily re-cast in a form suitable for both vector and parallel processing.

3.8.2 Algorithmic Details

Step 1 defines the geometry of the map. Usually this will be two-dimensional with the maximum number of 'clusters' equal to the maximum number of neurons, although in practice not every neuron need be assigned data, so the minimum number can be less. This is a very useful feature in classification in that there is no longer a one-to-one relationship between the number of neurons suggested by the user (usually in ignorance) and the best number of clusters or groups needed for the classification (usually unknown). For example, a 10 by 10 neuron array would contain 100 neurons around which clusters could develop. A neuron in this context is really a processing element. It consists of a vector of m data values (one for each variable of interest). So if the classification involves 20 variables and a 10 by 10 array of neurons is being used, then data storage is 10 by 10 by 20 numbers.

It is interesting to note that three-dimensional arrays might also be of interest in a geographical context. For example, in a data set in which regional variants of the key clusters might be expected, then the definition of topological neighborhood in step 5 of the basic algorithm might well include a geographical or locational component; i.e. the first two dimensions might be 100km square map coordinates with the third dimension being purely a dissimilarity space. Whether this is worthwhile is a matter for experimentation in particular contexts. One problem is that the different spaces need to be measurable in the same units or else a conversion function provided. This complicates things, for example, if space-time data are being used.

The one-dimensional neuron map array may also be of interest in that this forces a linear ordering on the data, so that position in the array will also reflect similarity to some degree. This might be useful. It would also be possible to imagine a continuous map that has no edges. Whether this offers any real benefits is uncertain. Finally, it seems to matter little as to whether the neurons are arranged on a grid or have a hexagonal geometry. The advantages of the latter mainly concern a better coverage of the two-dimensional space.

Step 2 merely describes how to start the process off. It might be best to use random realistically-scaled values, or even base the initial weights on randomly selected data values; or on random perturbations of variable means. It is probably not too critical provided they are not too extreme.

Step 3 is the first of the really useful sections of Kohonen maps. In a conventional cluster analysis algorithm, the data would be presented in sequential order. Each case having, implicitly, the same weighting. Here, however, it is much more useful to sample some data cases much more often, to reflect for example size of area or assumptions about the reliability of the data. This is extremely useful in a spatial data analysis context because small areas with small denominators often possess the most extreme data values. For example, it is much easier to obtain 100% unemployment in a small rural zone with 20 households than in the worst urban black spot with 200 households. The result of this can be catastrophic. The conventional classifier, treats all areas the same and thus tends to form cluster centroids based on the more extreme values (which are least reliable and meaningful) whilst regarding those areas with more reliable data values, to some extent, as outliers. This is the opposite of what you really want to happen. However, if you select cases randomly with probabilities depending on size then the impact of small number problems may now be greatly diminished and, hopefully, the quality of the final classification suitably enhanced.

Step 3 is also a useful place to add any relevant measurement noise to the data. Quite often in geography the variables being classified will possess different levels of measurement error. For example, in UK census analyses some variables are 100% coded others are 10% coded. Other data bases have missing values for some variables. It would be extremely useful to be able to take into account the varying reliability and accuracy of the data when classifying it. One simple way of achieving this goal is to randomize the values of the M data values by different amounts to reflect variable and even record specific levels of data noise. Provided there are sufficient training iterations there is no reason why this cannot be achieved, at least in theory. Again it is important with spatial data because noise, error, and spatially varying levels of reliability are all important attributes that need to be considered, rather than simply ignored. Neuroclassifiers offer considerable promise in this respect.

A key point to note is that the results reflect the probability density distribution from which the training cases were selected. This is very important because the key advantage offered here is that the user can use whatever probability density function is considered most appropriate and this need not be that defined in a completely arbitrary fashion by the available database. Biasing the training to give greater weight to the most reliable data values would seem to be extremely important given the nature of spatial information.

Step 4 is fairly simple and parallelisable. It determines which neurons 'win' the competition to represent the data case. There are various ways of measuring 'nearness' and 'similarity' and, depending on the form of data measurement, different measures might be appropriate. The program in Appendix II uses the sum of absolute differences as a measure of similarity.

Step 5 applies updating to the neuron weights in the 'neighborhood' of the winning neuron. The clever and adaptive behavior of the self-organizing map results from this part of the algorithm. The network is gradually 'tuned' to different inputs in an orderly fashion, almost as if a continuous mapping of the input space was formed over the network. The ordering and smoothing process is extremely subtle. The outcome is that different parts of the network become selectively sensitized to different input patterns. The size of the 'neighborhood' of the winning neuron is usually a function of time; that is, it decreases slowly as the training process proceeds.

The simplest updating algorithm is to define a block party based on distance from the winning neuron, so that all neurons within this critical distance update their weights. Kohonen (1984, 1989) defines this as:

$$m_i(t+1) = m_i(t) + \alpha(t)\,[x - m_i(t)] \qquad \text{for } i \in N_c(t) \qquad 1a$$

else

$$m_i(t+1) = m_i(t) \qquad 1b$$

where $m_i\ (t)$ is the weight vector for any neuron i which lies within the neighborhood set $N_c(t)$ of the best matching (winning) neuron c at iteration t; $m_i\ (t+1)$ is the updated weights for this neuron i; x is the vector of values for the data case; $N_c(t)$ is the neighborhood around winning neuron c at time t; and $\alpha(t)$ is a 'training' constant, typically $0 \leq \alpha(t) \leq 1$.

Some researchers prefer a slightly more sophisticated representation of the lateral inhibition process. Kohohen's (1984) brain inspired research argues in favor of a 'mexican-hat' function of lateral interaction. He writes "There is both anatomical and physiological evidence from the mammalian brains for the following types of lateral interaction to exist between cells: (i) short-range lateral excitation reaching up to a radius of 50-100μm (in primates); (ii) The excitatory area is surrounded by a penumbra of inhibitory action reaching up to a radius of 200 to 500μm; (ii) A weaker excitatory action surrounds the inhibitory penumbra and reaches up to a radius of several centimetres. The degree of lateral interaction is usually described as having the form of a Mexican hat" (p122-123).

One of the simplest ways of operationalizing this concept is to make the learning rate $\alpha(t)$ dependent also on distance (d) from the winning neuron (ie $\alpha(d,t)$). The previous updating algorithm is equivalent to a simple cut-off or step function and is achieved by setting $\alpha(d,t) = 1$ if a neuron is within N_c and $\alpha(d,t) = O$ otherwise. It seems that a precise representation of a Mexican hat function is not needed and that a useful, simple, alternative is a Gaussian function of distance from the winning neuron; see Ritter *et al* (1992). Muller and Reinhardt (1991) use the following function:

$$\alpha\,(d,t) = \exp\,(-\,d^2/2\beta\,(t)) \qquad 2$$

where d is the distance from the winning neuron, and $\beta(t)$ is the size of neighborhood at time t (previously N_C).

Step 6 is concerned with reducing the training parameters, slowly so that stability is achieved. It is important to note that the size of the neighborhood set $N_C(t)$ and the training constant $\alpha(t)$ both reduce slowly as a function of time (really the number of training iterations). Kohonen (1984) suggests that a proper choice for these parameters can best be determined by experience and that certain 'rules of thumb' can be invented. He adds ".. it may be useful to notice that there are two phases in the formation of maps that have a slightly different nature, viz., initial formation of the correct order, and final convergence of the map into asymptotic form. For good results, the latter phase may take 10 to 100 times as many steps as the former, whereby a low value of $\alpha(t)$ is used" (p133). Usually, the critical parameters are linearly decreasing functions of t. Kohonen (1989) uses the following scheme for two time periods (t_1, and t_2); for t_1 (the initial coarse structuring period):

$$\alpha(t) = k_1\,(1 - t/t_1) \text{ for } 0 \le t \le t_1 \qquad 3$$

and for t_2 (the fine structuring phase):

$$\alpha(t) = k_2\,(1 - t/t_2) \text{ for } t_1 \le t \le t_2 \qquad 4$$

where t_1 and t_2 depend on the dimensions of the neuron array and not on the number of variables, in Kohonen (1989):

$k_1 = 0.1$
$k_2 = 0.008$
$t_1 = 10\,000$
$t_2 = 90\,000$

and N_C decreased linearly from 12 to 1 during $0 \le t \le t_1$ and then remained constant.

In essence this gradual reduction in parameters is similar in principle to a cooling schedule in simulated annealing. The structure gradually emerges and crystallizes.

Step 7 suggests that the process needs to be repeated a fairly large number of times. Critical factors here are the number of neurons and possibly the size of training data set. It is probable that the best results will only emerge after several million training passes, particularly if noise in the data are also to be represented. It is also possible that computation constraints may well limit the full benefits of a Kohonen classifier until massively parallel processing becomes a practical proposition.

3.8.3 A Basic FORTRAN Program

There is really no problem in programming a Kohonen self- organizing map based classifier. However, having reviewed various published attempts at providing useful, usable, code then there is some justification for Appendix II. Basically, there is no need to litter the source code with misleading and irrelevant biologically inspired jargon; for example synopses, activation constants, sensory inputs, lateral correlations, synoptic change etc. Likewise some programs are seemingly far too complex; see for example, Freeman and Skapura (1991).

Appendix II should allow anyone interested to experiment with the technology. Whether it works well, brilliantly, or not at all is likely to be partially related to the user specified parameters, partly to the nature of the application, and partly the amount of computer power available.

3.8.4 Two Empirical Evaluations

The strongest justification for using a neuroclassifier is that it delivers 'better' results than conventional methods. 'Better' in a classification context is partly a matter of taxonometric performance which can be measured by a statistic of some kind, partly a matter of assessing the theoretical attractiveness of a more data sensitive and flexible technology, and partly the 'qualitative feel' that the user has about the classification results; for instance, are they easier to interpret, do they provide a 'clearer' description of the data. An appreciation of these subjective criteria require double blind testing and this is left as an interesting area for further research. Attention here is restricted to numeric measures.

Openshaw and Wymer (1990) reported the results of empirical tests performed on a small census data set of 120 areas and 45 variables. They compared the total percentage within cluster sums of squares for a variety of neuroclassifiers. The benchmark was a set of classifications produced by what is considered to be the best readily available conventional method (a K-Means procedure, see Openshaw (1983a)). In all cases the results of the Kohonen self-organizing map (SOM) based classifier was an improvement, albeit seemingly not a vast improvement. However, it should be noted that although similar cluster descriptions were obtained, the classifications were not the same, and even for this small data set there was a subjective preference for the SOM results. Also, the total percent within cluster sum of squares statistic based on 45 variables is a hard quantity to reduce, so even a slight change may be hard to achieve; for example by using a better conventional classification heuristic. It is also biased against the SOM because the Kohonen training was biased in favor of the more data reliable areas and the assessment of neuron performance was based on the sum of absolute errors and not the sum of errors squared used by the conventional classifier. Additionally, the conventional method used a principal components analysis that should also have helped.

A second test was based on the same 45 variables but for all 9,278 census wards in England and Wales (Table 3.1). These results are quite remarkable in that large improvements in performance are obtained. Furthermore, the results suggest that the geometry of the SOM is not too critical in terms of its impact on the results, once more

Table 3.1 Comparison of neural net results with a conventional classifier

Method used	Number of Clusters	% within cluster
Conventional	10	40.15
K	10	32.66
K:3*3	9	32.03
K:10*1	10	30.29
Conventional	20	34.21
K	20	26.82
K:5*4	20	24.93
K:20*1	20	25.14
Conventional	30	31.16
K	30	23.12
K:10*3	30	21.76
K:30*1	30	22.00
Conventional	40	27.84
K	40	21.35
K:8*5	40	20.23
K:40*1	40	20.08

Notes:
The conventional classification was based on a K-means procedure
K is a simple competative net
K:a*b is a SOM with a by b neurons in it

than a relatively small number of clusters are considered. A wider comparison with other types of neuroclassifier is contained in Openshaw, Wymer, and Charlton (1993). The latter suggests that the SOM is the best classifier.

The cost of obtaining these improved results is, or was, a reliance on a Cray XMP supercomputer. Run times of 30 minutes CPU time were typical. However, this is today quite modest in that it translates into less than three hours on a fast Sun workstation; such has been the speed-up in computer hardware performance over the last two years.

The importance of proving 'good performance' however cannot be underestimated. We need to discover when and when not to use neuroclassifiers. Certainly, if we seek to do no more than what existing methods can deliver then the benefits of using neuroclassifiers may be minimal (i.e. perhaps a slightly better optimization) and probably not worth the effort. If, however, we seek to produce less assumption dependent classifications then (a) there is no longer any conventional alternative and (b) we may indeed obtain significantly improved results. We also need to experiment with a range of architectures. The principal benefits would appear to be an ability to produce more geographic data sensitive and less assumption dependent classifications.

3.9 Conclusions

This chapter has made the case for at least considering the application of grandiosely sounding neuroclassifiers when seeking to analyze spatial data. The preferred method is the self-organizing 'map' due to Kohonen (1984). The benefits to the geographer can be summarized as follows: fewer assumptions that matter, simple programming suitable for both vector and massively parallel hardware, technology that can readily be made to become spatial data sensitive, and the prospect of significantly improved spatial classifications. It could also be added that these methods are nonlinear, based on emergent bottom-up behavior, do not force the data into a predefined structure, and appear to yield results that have a more natural feel about them. This new technology was, and still is, fashionable because of the falling out of computing and its underlying biological inspiration. Despite its seemingly simple structure, the results and the behavior of the processing elements can be highly complex.

Self-organizing, selfish processing elements (or neurons) provide new, more natural, and seemingly 'better' ways of classifying spatial data. However, it is still early days so the onus rests with the researcher to demonstrate that the effort has been worthwhile. Preliminary results suggests that in a geographical context, at least, it is certainly worth making the effort.

References

Amari, S-I. (1972) "Learning Patterns and Pattern Sequences by Self-Organizing Nets of Threshold Elements" IEEE Transactions on Computers 21, 1197-1206.

Beale, R., Jackson, T. (1990) "Neural Computing: An Introduction Adam Hilger, Bristol.

Carpenter, G., Grossberg, S. (1987a) "Invariant Pattern Recognition and Recall by an Attentive Self-Organizing ART Architecture in a Non-Stationary World", Proc. of IEEE First Int. Conference on Neural Networks Vol II, IEEE, San Diego, 737-746.

Carpenter, G. Grossberg, S. (1987b) "ART2: Self-Organization of Stable Category Recognition Codes for Analog Input Patterns", Applied Optics 26, 4919-4930.

Carpenter, G.A., Grossberg, S., Reynolds, J.H. (1991) "A Self-Organizing ARTMAP Neural Architecture for Supervised Learning and Pattern Recognition", in R.J. Mammone, Y. Zeevi, Neural Networks, Theory and Applications, Academic Press, Boston, 43-80.

Caudill, M., Butler, C. (1992) "Understanding Neural Networks: Volumes 1 and 2", Bradford, MIT Press.

Freeman T A., Skapura, DM. (1991) "Neural Networks: Algorithms, Applications and Programming Techniques", Addison-Wesley, Massachusetts.

Grossberg, S. (1972) "Neural Expectation: Cerebellar and Retinal Analogues of Cells Fired by Unlearnable and Learnable Pattern Classes", Kybernetik 10, 49-57.

Grossberg, S. (1976) "Adaptive Pattern Classification and Universal Recoding II: Feedback, Oscillation, Olfaction and Illusions", Biological Cybernetics 23, 187-207.

Grossberg, S. (1988) "Neural Networks and Natural Intelligence", MIT Press, Cambridge, Massachusetts.

Hopfield, J. (1982) "Neural Networks and Physical Systems with Emergent Collective Computational Abilities", Proc. of the National Academy of Sciences 70, 2554-2558.

Hopfield, J. (1984) "Neurons with Graded Response have Collective Computational Properties Like Those of Two State Neurons", Proc. of the National Academy of Sciences, 81, 3088-3092.

Kanerva, P. (1988) "Sparse Distributed Memory", MIT Press, Cambridge, Massachusetts.

Kohonen, T. (1977) "Associative Memory - a System Theoretical Approach", Springer-Verlag, New York.

Kohonen, T. (1984) "Self-Organization and Associative Memory", Springer-Verlag, Berlin.

Kohonen, T. (1989) "Speech Recognition Based on Topology Preserving Neural Maps", in I. Aleksander (ed), Neural Computing Architectures, Chapman and Hall, Andover.

Kosko, B. (1988) "Bidirectional Associative Memories", IEEE Transactions on Systems, Man, and Cybernetics, 18, 42-60.

Malsburg, C. (1973) "Self-Organization of Orientation Sensitive Cells in the Striate Cortex!", Kybernetik 14, 85-100.

Muller, B., Reinhardt, T. (1991) "Neural Networks", Springer-Verlag, Berlin.

Openshaw, S. (1976) "A Regionalization Procedure for a Comparative Regional Taxonomy of the UK", Area 8, 149-152.

Openshaw, S. (1983) "Multivariate Analysis of Census Data: The Classification of Areas", in D. Rhind (ed), A Census Users Handbook, Methuen, London, 243-264.

Openshaw, S. (1989) "Making Geodemographics More Sophisticated", Journal of the Market Research Society 31, 111-131.

Openshaw, S. (1991) "A View on the GIS Crisis in Geography or Using GIS to Put Humpty-Dumpty Back Together Again", Env. and Planning A, 23, 621-628.

Openshaw, S. (1992) "Some Suggestions Concerning the Development of AI Tools for Spatial Modelling and Analysis in GIS", Annals of Regional Science 26, 35-51.

Openshaw, S. (1992b) "A Review of the Opportunities and Problems in Applying Neurocomputing Methods to Marketing Applications", Journal of Target Marketing, Measurement and Analysis 1, 170-186.

Openshaw, S. (1992b) "Further Thoughts on Geography and GIS: A Reply", Env. and Planning A 24, 463-466.

Openshaw, S. (1993a) "Modelling Spatial Interaction Using a Neural Net", in M.M. Fischer and P. Nijkamp (eds), GIS, Spatial Modelling and Policy Evaluation, Spring-Verlag, Berlin, 147-164.

Openshaw, S. (1993b) "Special Classifications", in B. Leventhal (ed), An Introductory Guide to the 1991 Census, NTC Publications, London, 69-81.

Openshaw, S., Wymer, C. (1991) "A Neural Net Classifier for Handling Census Data", in F. Murtagh (ed), Neural Networks for Statistical and Economic Data, Munotech, Dublin.

Openshaw, S., Wymer, C., Charlton, M.E. (1993) "A Comparative Evaluation of Three Neuroclassifiers of Census Data" (forthcoming).

Ritter, H., Martinetz, T., Schulten, K. (1992) "Neural Computation and Self-Organizing Maps, Addison-Wesley, Massachusetts.

Rosenblatt, F. (1962) "Principles of Neurodynamics", Spartan Books, Washington.

Rumelhart, D. Zipster, D. (1985) "Feature Discovery by Competitive Learning".

Sokal, R.R., Sneath, P.H.A. (1963) "Principles of Numerical Taxonomy", Freeman, London.

Simpson, P.K. (1990) "Artificial Neural Systems", Pergamon Press, Oxford.

Spence, N.A., Taylor, P.J. (1970) "Quantitative Methods in Regional Taxonomy", C. Board (ed), Progress in Geography 2, Arnold, London, 1-63.

Wilson, A.G. (1974) "Urban and Regional Models in Geography and Planning", Wiley, London.

Stan Openshaw, School of Geography, University of Leeds, Leeds LS2 9JT, United Kingdom

Chapter Four

SELF ORGANIZING MAPS -- APPLICATION TO CENSUS DATA

Kevin Winter and Bruce C. Hewitson

In the previous chapter, Stan Openshaw provides a lucid overview of the classification of spatial data using neural-nets. In this chapter we follow with an example that uses one form of neural classification, namely the Kohonen Self Organizing Map (SOM). In this case, the SOM is used to organize demographic data gathered from census information in order to investigate population groupings and their spatial distribution. In other words, to perform an unsupervised classification, or a mapping of a ten dimensional input to a simple two dimensional surface, or in the full jargon, a non-linear projection onto two dimensions of the probability density function. As this is an example of the techniques already discussed, we assume that we need not repeat the theory and practice of the preceding chapter.

4.0 South African Census Records

The data contain some unusual structural characteristics as it is taken from the 1991 South Africa census record and demonstrates the unique social structures of the country. South African society has a history of racial stratification brought about by a 'divide and rule' concept used to the advantage of the ruling regime. The society is characterized by racial separation from the apartheid era, and in particular has disadvantaged the 'black' and 'colored' groups in terms of their access to fundamental resources such as land, education and opportunities for work. Since the data represents many racial terms, such as 'black', 'whites', 'coloreds', etc., the use of these terms in this chapter is not intended to indicate support of racial bias. A further point of clarification needs to be made with respect to the use of the term 'coloreds'. This term refers to a distinct group within the South African society that arose from the inter-marriage and inter-breeding of the indigenous population with the early colonial settlers. The group has developed a distinct culture of its own and it plays a significant role in the activities of the region.

The data represent a random subset of people from 21 districts in the Western Cape region of South Africa. The data represent an unadjusted enumerated population

B. C. Hewitson and R. G. Crane (eds.), Neural Nets: Applications in Geography, 71–77.

of approximately 2.1 million. A weighting factor, intended to correct census areas that were either sampled or inaccesible, has not been applied to the data and, therefore, the data may not accurately reflect the entire population living in certain areas. The census districts cover the spectrum of urban and rural areas, such as the central city region of Cape Town, the upper, middle and low class suburbs, informal settlements, rural agricultural districts, and smaller coastal and mountain settlements. The population spans all economic sectors and ranges from the upper class westernized sector of the population to the poorer sectors which are dominated by the 'black' and 'colored' population groups in particular. From the census data ten demographic variables were selected as shown in Table 4.1.

Table 4.1: Demographic Variables Extracted from the 1991 South African Census and Used in the Neural Net Classification.

Variable	Numerical Range	Comment
Sex	1/2	Male/Female
Age	1-99	Number of years
Race - White	0/1	0 = No 1 = Yes
Race - Colored	0/1	0 = No 1 = Yes
Race - Asian	0/1	0 = No 1 = Yes
Race - Black	0/1	0 = No 1 = Yes
Education Status	0-20	Equates to increasing education
Income	0-21	Equates to increasing income
Employed/Unemployed	0/1	0 = Employed 1 = Unemployed
Marital Status	0/1	0 = Unmarried 1 = Married

4.1 Net Classification

From the full census data for the region, 107,604 records were randomly selected from all 21 districts to use as the training data for the SOM, and a larger subset of 429,469 records was randomly selected for mapping a larger sample once the SOM organization was established. The application of the SOM is in the form of the standard Kohonen procedure as described in steps 1-7 in the previous chapter, and is outlined here in the same order as the steps in the previous chapter.

Step 1: Three separate nets are actually defined, all with a hexagonal topology, but of different dimensions. The dimensions used are 8 by 5 nodes, 15 by 12 nodes, and 24 by 18 nodes. The intention is to demonstrate how the established features or structure becomes more detailed with increasing numbers of nodes.

Step 2: Each node has ten weights associated with the ten variables of the data. These are initialized to random normalized values.

Step 3: Not applied.

Step 4 - 7: The training data are applied in a random order in a two stage process. In the first stage, the data are applied to the net with a large neighborhood function and a moderate sized learning parameter. The intention is to organize the map into the gross features of the data. A second pass is then performed where the entire 107,604 records are applied numerous times with a small learning parameter and a neighborhood function that only updates the most immediate neighborhood nodes. This is the 'fine-tuning' of the net, allowing it to focus on the more detailed aspects of the classification.

At this point the net is deemed to have determined the optimum mapping function from the input space onto the two dimensional node array. The larger subset of data, 429,469 records, is now passed to the net in order to obtain a dimensional mapping of each record, for example record 1 being mapped to node {2,5}, etc. This information, in conjunction with the extracted weights for each node in the net, forms the basis for the interpretation.

For each node there are ten weights associated with each of the input variables. Thus if one considers, for example, the first weight of a nodes' vector, the set of these may be visualized by generating an image of the net dimensions where the value of each 'pixel' in the image is determined by the first weight value from the vector. In the case of the first net, with the 8 by 5 node dimensions, an 8 by 5 'pixel' image is produced, or in the case of the 24 by 18 net, a 24 by 18 'pixel' image. The image indicates how the first weight of all the node weight vectors has been organized across the nodes. Since the node weight vectors are targets to which the inputs are mapped, the first element of the weight vector represents the first variable of the input, in this case the sex structure of the population. Consequently, by generating images for each element of the weight vectors, in this case a total of ten, visual pictures are created of the mapping function of the net for each input variable. In the case of the first element of the weight vectors, the image indicates a continuous 'mapping surface' for the sex structure of the population. For each of the three nets, a set of ten images was created as shown in Figure 4.1.

If one selects one of the Figures, 4.1a, 4.1b, or 4.1c, the interpretation can take two perspectives. Either by considering the mapped surface of a particular variable, or through selecting a location of interest in one image, and using the corresponding location in the other images to obtain the spectrum of demographic characteristics. Note that each image in Figure 4.1 contains the same structural pattern, although mirrored in 4.1a with respect to 4.1b and 4.1c. The high resolution net with 24 by 18 nodes provides 432 mapping points across the 2 dimensional surface, and reveals far more of the finer structure than does the 8 by 5 node net. In comparison, the medium resolution 15 by 12 net establishes the same features with the exception of a few finer featured details. The major portion of the population structure is captured at the 15 by 12 resolution, and hence further discussion will focus on Figure 4.1b.

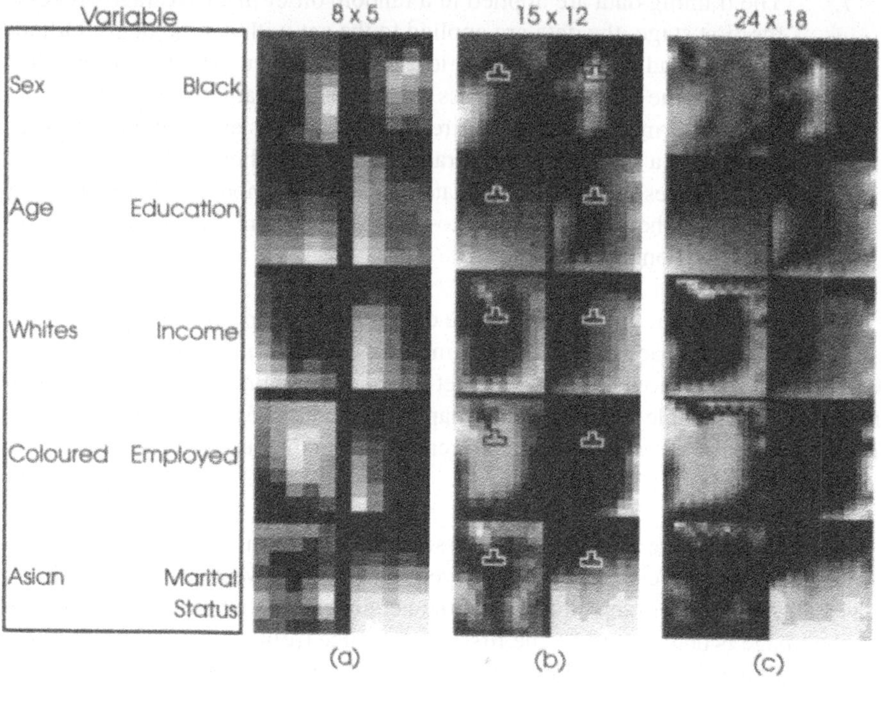

Figure 4.1: A visual representation of the population variables as mapped by the net. See text for further discussion.

4.2 Interpretation of the Mapping Surface

It will be helpful for those unfamiliar with the South African socio-economic context to make some generalization about the overall patterns that have emerged, before particular nodes or groups of nodes are examined in detail. In the first instance, the population sex structure in Figure 1b indicates a society which is well stratified by male and female. Males are represented by the dark tones in the image, while females are represented by the lighter tones. The age structure in Figure 4.1 shows a gradual change from dark tones, indicating the predominantly youthful population, to lighter tones which indicate the older population groups. South Africa is a developing country, and like many other countries in this stage of development, it is characterized by rapid population growth, with the fastest growth rate being experienced in the 'black' population group (Grobbelaar, 1990). The next four images in the series represent the various cultural groups, which are classified as 'race' groups in South African society. The stratification of these groups is readily noticeable in these images. The 'white' group is dominant in the lower right hand side of the image but is noticeably absent to the center of the image. By contrast, the 'colored' group dominates the center of the next image. The third population group is that of the Asian group, people who are largely of Indian descent. There is a relatively small representation of Asian communities in the Western Cape region and this may explain the diffuse distribution of nodes seen in the image. By contrast, the 'black' population demonstrates, along with the 'colored' group, a powerful image of a society which has been divided by colonial rule and Apartheid legislation. The 'black' population group dominates the center of the image, clearly dissimilar to the 'white' and Asian groups.

The low level of education is particularly noticeable in the case of the 'black' and 'colored' groups, whilst the attainment of higher education levels is reflected by the 'white' and older members of the population. Furthermore, the lower education levels and lower income levels relate well to the employment image. Finally, the lighter tones shown by the marital status image describes those members of the poplulation who are either married or living together, whilst the darker tones indicate single marital status.

4.3 Interpretation of Regions in the Mapping--The 'Black' Population

Four nodes have been highlighted in Figure 4.1b. These nodes were selected as they demonstrated a high clustering of the 'black' population group variable. The socio-economic characteristics of this population group can be interpreted by examining the corresponding pixels in the images of the other variables. A brief analysis of each reveals the following information:

- the group is equally represented by males and females;
- the group is dominated by a youthful population;
- there is little or no relationship between the 'white' and Asian groups and the 'black' population group. However, there is a much stronger relationship between the 'colored' and 'black' groups;
- the 'black' group has a relatively low level of education and a corresponding low level of income;
- the high level of employment is, however, unrealistic of the South African

situation. This anomaly may be explained by the fact that in the census, the largest urban 'black' community was not enumerated by household, but only sampled.
Furthermore, the data used for this exercise does not include the census weighting factor described earlier. Alternatively, as discussed in the previous chapter, deliberate over-representation of this group during the training of the net may result in better differentiation of the clustering; and
- the relatively youthful population group is predominantly unmarried.

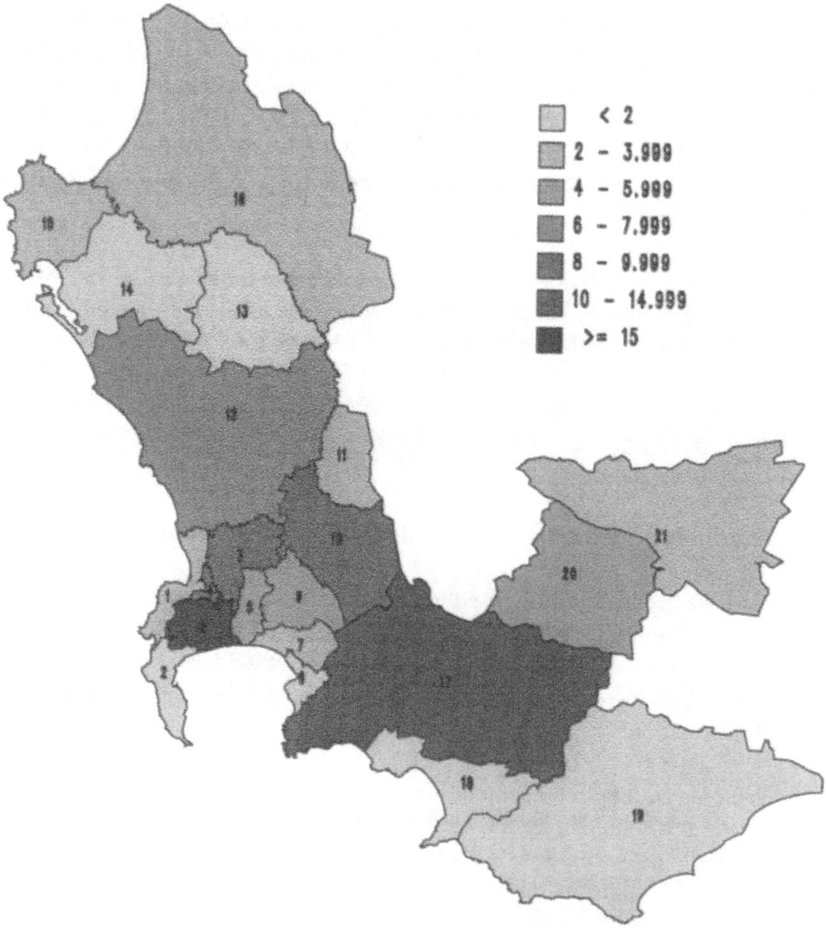

Figure 4.2: The percentage number of cases in each district that are identified as belonging to the nodes highlighted in Figure 4.1b.

4.4 Spatial Distribution of the Mapping

As all records carry spatial location information that relates to census districts, the people mapped to the four selected nodes may be spatially represented on a map of the census districts of the Western Cape region (Figure 4.2). The largest proportion of the black population is situated in district 6, which is also the most densely populated district in the Western Cape. This district is characterized by a range of urban, surburban and informal housing. The map also indicates the stratification of population distribution by 'race'. For example, with the exception of disctrict 6, the black population lives in rural areas on the periphery of the Cape Town Metropolitan area (district 1). Thus it is demonstrated that the political history of South Africa has given rise to a very strong relationship between the actual spatial distribution of the population and the mapping of people to nodes in the neural net.

4.5 Conclusions

The interpretation of large multidimensional data sets, and social data in particular, have been enhanced by the neural net application. The SOM technique, in a non-linear and unsupervised manner, has demonstrated an ability to reveal some of the underlying social structures of the population. While this represents a generalization of a highly complex data set, the net nevertheless provides a powerful means of describing the social data. A twofold challange remains, to place the correct interpretation on the data, in particular with respect to the more subtle features, and then most importantly, to correctly use the information for furthering the development of the region.

References

Grobbelaar, J. (1990) "Forecasts of South African Population for the Period 1985 - 2020", Occasional Paper No 17, Institute for Futures Research, Stellenbosch, SA., Changing South Africa, OUP, Cape Town.

Kevin Winter, Department of Environmental and Geographic Sciences, University of Cape Town, Private Bag, Rondebosch, 7700, South Africa.

Bruce C. Hewitson, Department of Environmental and Geographic Sciences, University of Cape Town, Private Bag, Rondebosch, 7700, South Africa.

Chapter Five

PREDICTING SNOWFALL FROM SYNOPTIC CIRCULATION: A COMPARISON OF LINEAR REGRESSION AND NEURAL NETWORK METHODOLOGIES

David L. McGinnis

5.0 Introduction

Much of the current debate regarding global warming and the regional manifestation of climate change relates to the ability of General Circulation Models (GCMs) to adequately represent modern climate features. Although the GCMs represent large-scale features well, their ability to model regional or local climate remains highly suspect. Important developments for the analysis of regional climate-change include statistical methodologies that translate between the large-scale and local-scale extremes. Traditionally, such methodologies have been limited to linear relationships; however, neural networks provide a new way to accomplish the same goal, with the added benefit of addressing the non-linear relationships that are characteristic of some climatic fields.

Hewitson and Crane provide examples of both approaches: predicting temperature distributions from atmospheric circulation using stepwise linear regression (Hewitson and Crane, 1992a), and using neural nets to demonstrate relationships between atmospheric circulation and regional precipitation (Hewitson and Crane, 1992b; see also Chapter Seven of this volume). The two studies were carried out for very different climate regimes. The first paper examined linear relationships between the atmospheric circulation and the temperature distribution over the United States, while the second paper used neural nets to examine relationships between circulation and rainfall in tropical Mexico--a location where precipitation is very discontinuous in time and is dominated by convective processes. The neural net performed well under these conditions (see Chapter Seven), but there was no direct comparison of neural nets versus more traditional linear procedures.

The goal of this chapter is a direct comparison of linear-regression methods with neural networks applied to an identical environmental data set. The scientific objective is to determine relationships between snowfall and atmospheric circulation that influence the spatial and temporal distributions of snow cover in the Upper Colorado Basin of the United States. The Colorado River supplies most of the water for domestic, agricultural, energy, and recreational use in the southwestern states and parts of Mexico. With

B. C. Hewitson and R. G. Crane (eds.), Neural Nets: Applications in Geography, 79–99.

approximately 70% of Colorado River water beginning as high-altitude snowpack, the Colorado Basin provides an excellent case study to develop a methodology relating synoptic circulation to snowfall and, therefore, to water availability in the arid regions served by the Colorado River. This chapter discusses the data preparation used for the comparison, and also reviews possible circulation-snowfall relationships using more 'traditional' stepwise linear regression techniques. The improvements that result from applying neural nets to the same data set are then examined, and some conclusions regarding scale-translation methodologies drawn.

5.1 Data Preparation and Methodology

Data for this study come from two major sources. The atmospheric circulation is determined from a set of observational products supplied by the National Center for Atmospheric Research (Jenne, 1975). These National Meteorological Center (NMC) data are derived from station observations using a dynamic model (see Cressman, 1959) and are twice-daily gridded observations placed on a polar stereographic projection; therefore, the grid points are more or less equidistant across the spatial domain. The 700 mb data were chosen because the mean height of the 700 mb field approximates the average elevation in the study area; the 700 mb level is generally in the neighborhood of 3000 meters and much of the Upper Colorado Plateau lies between 3000 and 4000 meters. The study area chosen is one that captures the major atmospheric flow patterns affecting the Colorado Plateau. Hence, the NMC 700 mb grid window used extends from 125°W to 100°W longitude and 30°N to 50°N latitude and includes 44 gridpoints (Figure 5.1). A time series (daily observations at 1200 Z) from 1981 to 1989 is selected in order to match the time domain of the available snowfall data.

The source for the daily snowfall measurements is the Soil Conservation Service West Technical Center (SCS-WTC) Snow Telemetry (SNOTEL) historic data base. The SNOTEL sites are automated sensors designed to provide daily measurements of snow-water equivalence (SWE), precipitation, minimum temperature, and maximum temperature (temperature readings are determined at approximately six feet above ground level). For this study, only the SWE data are used. These measurements are obtained from the pressure exerted by the snowpack on a flexible pillow approximately eight feet in diameter; the pressure reading is then converted to a value for SWE. These SNOTEL sites are distributed within each sub-basin at a variety of elevations and aspects and, where possible, are placed in forest clearings where drifting is unlikely. Each evening, these automated sensor readings are collected by the WTC, and archived for current water-year use and historical analysis. There are now more than 100 SNOTEL sites on the Colorado Plateau; however, this study uses only the 54 sites having the longest data record and highest data quality. Thus, these 54 SNOTEL sites, distributed over various elevations, aspects, and drainage basins, are taken here to represent the spatial distribution of the regional snowfall (Figure 5.2).

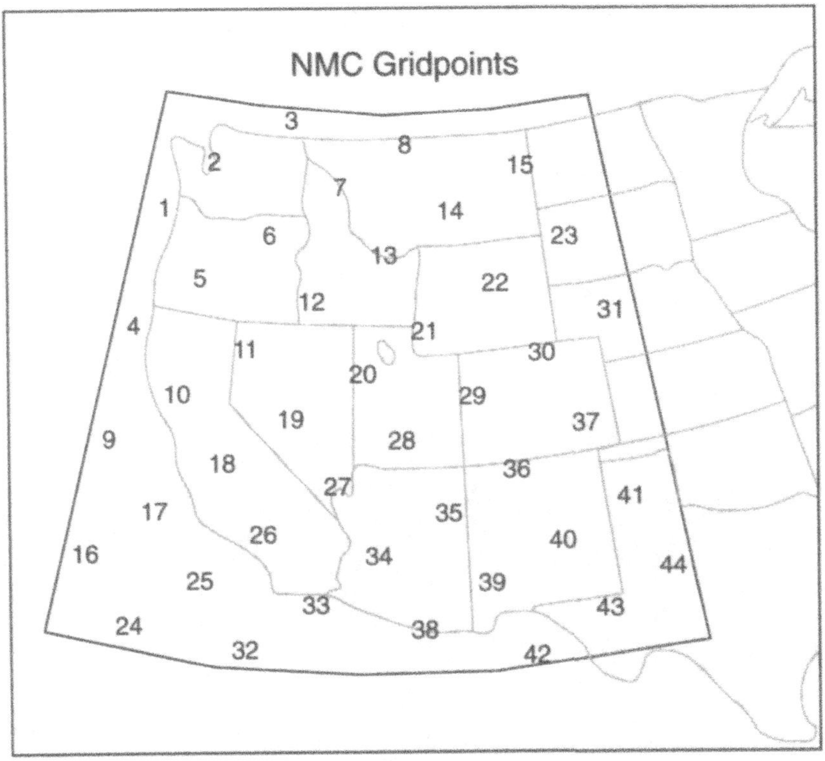

Figure 5.1: NMC gridpoints used to model 700 mb circulation.

5.2 Principal Component Analysis--700 mb Data

Preparation of the data for analysis involves several steps. An S-mode rotated Principal Component Analysis (PCA) is applied to the 700 mb time series as a means of data reduction. The S-mode analysis produces orthogonal patterns of common variance, where the spatial variance is represented by the component loadings and the PCA scores represent the component time series. Thus, the snowfall on each day is represented by some combination of loading patterns weighted by their respective component scores. The input matrix for the PCA has 44 gridpoint variables forming the columns (*m*) and 2644 days forming the rows (*n*). The principal components are extracted from the (*m* x *m*) correlation matrix where the components are linear combinations of the original pressure variables, and represent abstractions describing cohesive patterns of variance in the correlation matrix. These are orthogonally rotated using the varimax rotation to

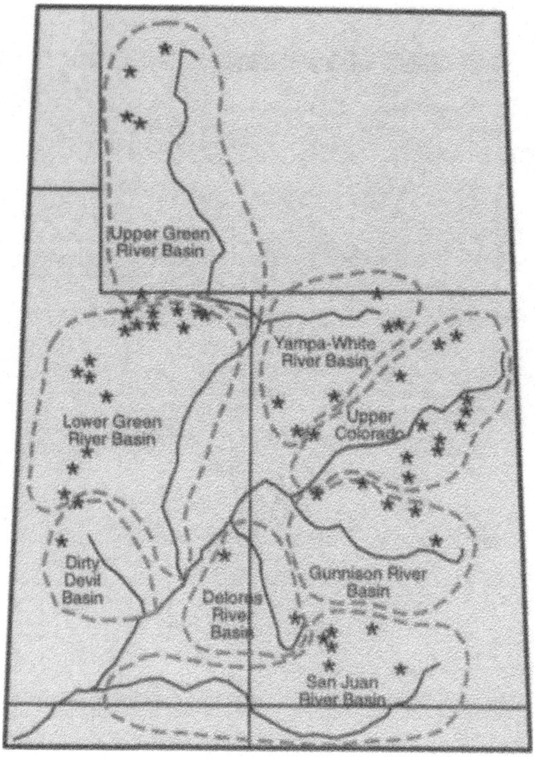

Figure 5.2: SNOTEL site locations.

maximize the explanatory power of each component. When the eigenvalues are determined, any value less than 1.0 is likely to be random noise in the data--the larger the eigenvalue, the more information it carries. In this analysis, the first eigenvalue was 33.28 explaining 76% of the data set variance, the second eigenvalue equalled 4.62 explaining another 11% of the variance, and the third eigenvalue was 3.22 adding another 7% to the explained variance. Further eigenvalues were all below 1.0 and are not considered in the analysis. The PCA resulted in three valid components that explain a total of 93% of the variance in the 700 mb time series. When rotated, these components explain 40.6%, 27.1% and 25.8% of the total variance respectively. The spatial patterns of these components, as described by their loading patterns, are shown in Figure 5.3. These loadings may be thought of as correlations with the eigenvector such that where the 700 mb time series at a grid point is most like the time series of the component, the loading will be high. Therefore, in component I the variance pattern covers the southeast portion of the study region--most likely demonstrating the influence of synoptic systems resulting from cyclogenesis in the lee of the Rockies. The second

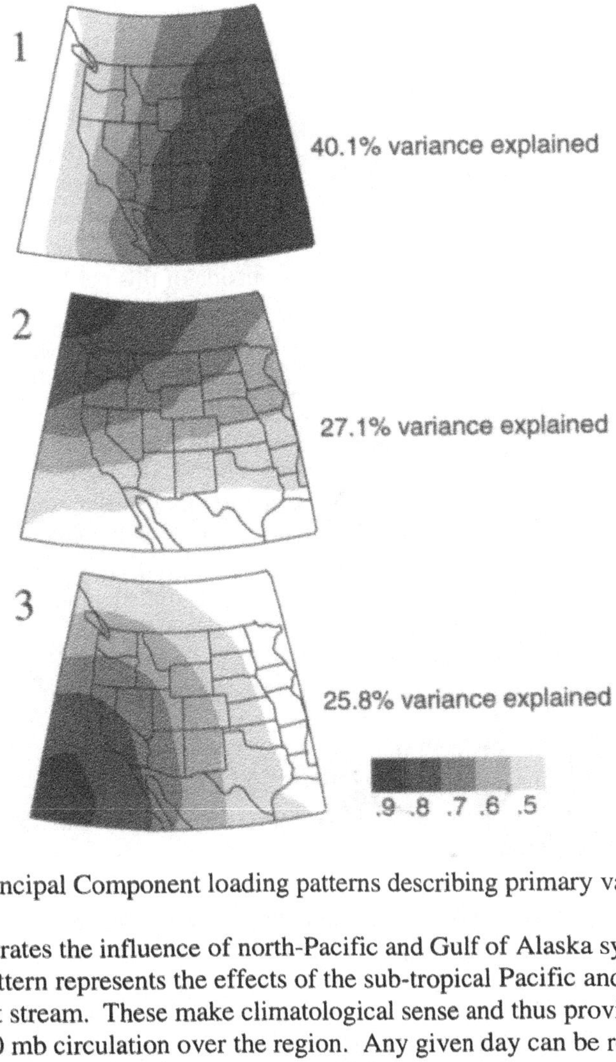

Figure 5.3: Principal Component loading patterns describing primary variance regions.

pattern demonstrates the influence of north-Pacific and Gulf of Alaska synoptic systems, and the third pattern represents the effects of the sub-tropical Pacific and the southern branch of the jet stream. These make climatological sense and thus provide a 'simplified' index of the 700 mb circulation over the region. Any given day can be represented as a combination of these three variance patterns and the component score values calculated by combining the original data time series with a score coefficient matrix determined by the PCA procedure. The score values become the input data to the linear regression and neural net algorithms used to predict snowfall.

5.3 SNOTEL Data Preparation

The SNOTEL daily snowfall values received from the WTC are presented as cumulative snow accumulation at each site for a given snow year. The study period chosen is a compromise between the length of the available record and the number of SNOTEL sites in operation. Fifty-four sites in the Colorado Basin extend back to 1981; the 1981-1989 time period is used for the present analysis. The first step in pre-processing the SNOTEL data is to convert the cumulative values into daily snowfall amounts, and to check for missing data days, and days with unreasonable snowfall amounts. For this analysis, snowmelt days are assumed to be 'no snow' days despite the possibility that liquid precipitation may have occurred. The snow year is defined as 1 October to 31 May. While some snow may fall outside of this period, this is the main snow accumulation period determined by the snow experts at SCS. Each of the 54 SNOTEL sites was individually analyzed and any snowfall events greater than five standard deviations from the mean at that station were deleted from the time series. For example, one station reported a one-day snow event of 25.80 inches of snow-water equivalence. By normal estimating methods, this would have been a single-day event of well over 200 inches of snow--not a likely event! The 54 SNOTEL sites in the study area had a average snowfall per year of 15.2 inches of snow-water equivalence; the range of average station snowfall was 5.9 to 37.2 inches of snow-water equivalence.

Finally, for each methodology used to create the statistical inversions between atmospheric circulation and snowfall, the appropriate data structure was determined that best suited the analysis. These data structures are discussed in the sections below; however, in all cases, the 700 mb and the SNOTEL data records were merged such that only days common to both data sets are included for analysis. No days are utilized with interpolated or missing data from either data source. The data are organized such that the snowfall is predicted from the 700 mb PCA scores. For each methodology (linear regression or neural net), several data configurations were tested but only the best results from each are described below.

After removing spurious data and isolating those stations with coherent records, a five-day moving average filter is applied to the data set for the linear regression analysis. To determine the appropriate smoothing time span, the snowfall events were characterized based on the length of time for a single snowfall event--defined as continuous days with snowfall recorded (Figure 5.4). It is apparent that the vast majority of snow events last only one day; however, multi-day events tend to produce the more significant snowfalls, especially in the early and late snow season. Most of these multi-day events are of 2-5 days in duration, hence the 5-day smoothing is used (other smoothing periods were tested including raw daily snowfall, but the 5-day smoothing performed best in the regression analysis). The daily snowfall and the 700mb index are averaged over the same five-day period.

Snowfall over the Colorado Plateau occurs due to a variety of factors (Changnon, *et al.*, 1990) but this study assumes that the majority of snowfall occurs as a direct result of atmospheric circulation patterns. When the daily snowfall values are

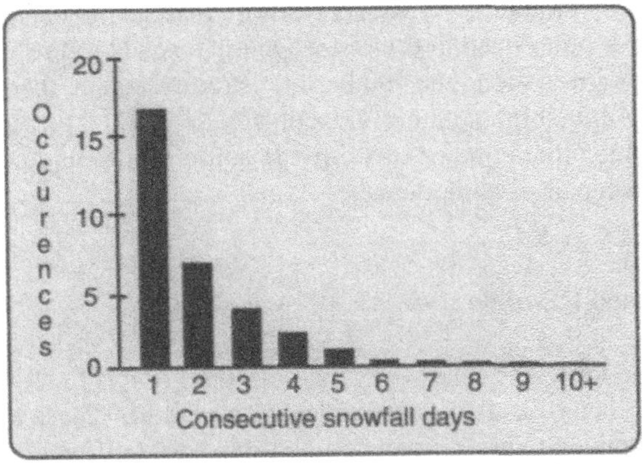

Figure 5.4: Average-annual consecutive-day snowfall events.

smoothed, the synoptic system controls are enhanced in the signal because the non-synoptic controls are minimized in comparison to the large-scale influences. Additionally, the initial stepwise regression procedures are performed for each individual SNOTEL station included in the study so each station has its own unique regression model. These regression models best emulate the linear relationships between the synoptic circulation and snowfall at that station, thus translating from large-scale atmospheric circulation to small-scale hydro-climatic variable prediction. The 5-day smoothed data are used only in the linear regression analysis; in the non-linear neural network analyses, the 5-day smoothing did not perform as well as the raw daily snowfall values, thus a different method of capturing the synoptic circulation was used. Consequently, the pre-processing of the data sets was not identical for each analysis but was, in fact, modified to provide the optimum results for each approach.

5.4 Stepwise Multiple Regression Analyses

Stepwise multiple regression analysis is a standard tool in climatology, employed when a variety of independent variables might affect the prediction of a dependent variable. In this analysis, a polynomial regression is used, where the independent variables are the scores from the PCA of the 700 mb time series (three components), their squares and their cross-products. Thus, a total of nine independent variables are used to characterize the synoptic circulation conditions over the five-day period; the dependent variable is the snowfall at each individual station. The PCA scores represent the similarity between the variance pattern on that day and the loadings patterns (Figure 5.3). If, for example, the five-day period is characterized by a cyclone forming in the lee of the Rockies (pattern for component I) followed by high pressure originating to the northwest (component II) then the average scores for components I and II will both be high on that day. On the other hand, if the same cyclogenesis is followed by a low pressure system being carried along the southern jet stream branch, then the scores on components I and III will be high with component II being low. The cross products of each component with the other components provide some representation of the interaction between each pair of components.

5.5 Five-Day Smoothed Results

The highest and lowest r^2 values for any individual SNOTEL station were 0.57 and 0.14 respectively. The overall 54-site average r^2 was 0.34. These represent the highest correlations obtained overall, when comparing several different smoothing periods. Six day or longer smoothing did not result in any appreciable increase in correlation coefficients; the lowest correlations were obtained with raw daily values (no smoothing), where r^2 values ranged from 0.08 to 0.48 with a 54-site mean of 0.25. These values are not particularly high but they do show that the atmospheric circulation and snowfall have between 14% and 57% of their variance in common, suggesting that there is some relationship between the two. The spatial distribution of the r^2 values (Figure 5.5) also demonstrates some regional autocorrelation where the r^2 values are grouped by river drainage basins. The SNOTEL sites close to one another tend to have approximately the same amount of explained variance and there are larger differences in explained variance from one sub-region to another. The highest-elevation sub-regions of the Upper Colorado River, when averaged together, have an r^2 of only 0.26. However, this region, supplies more than a quarter (26.3%) of the total runoff to the Colorado River based on the last 30 years of record. Unfortunately, the stepwise regression relationship is strongest in the Dirty Devil and the Lower Green River drainage basins, which supply the least water to the Colorado River flow. Thus, in this highly variable terrain, the linear methodologies do not provide a strong relationship between atmospheric circulation and snowfall for those basins where the snowfall is most critical for the Colorado River runoff.

The regression models are used to predict snowfall at each site. As a further test of the circulation's control on snowfall, these predictions can be compared to the actual snow that fell on that day and the residuals from the regression model predictions calculated (note: this is done with the same time series used to derive the models--not an independent data set). These residuals seldom range above zero and normally exhibit an underprediction of snowfall. The residuals for each day were calculated and averaged throughout the time period so a yearly pattern would emerge. The stations were then all averaged so the seasonal pattern of the regression model predictions to actual snowfall was determined (Figure 5.6). The regression models underpredict most severely early and late in the snow season due to model's inability to predict the heavy-wet snowfalls of these periods. During the middle of the snow season the models perform better, but still

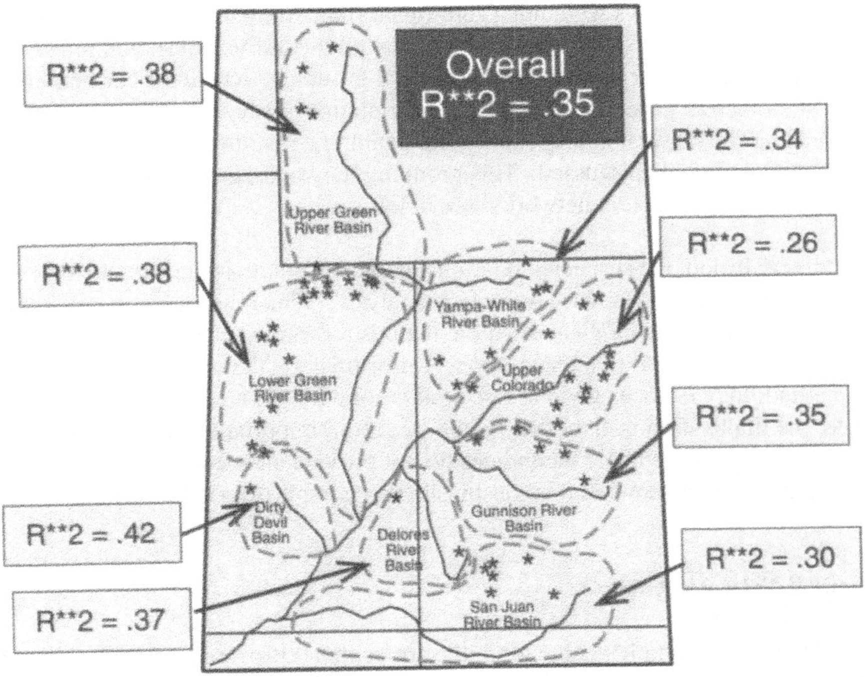

Figure 5.5: r^2 values averaged by Colorado River sub-basins.

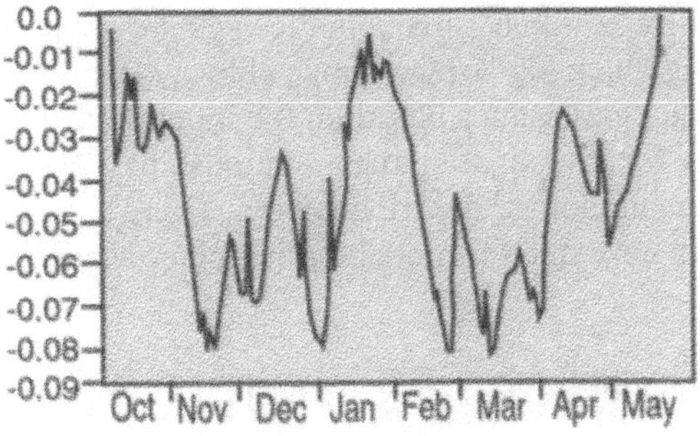

Figure 5.6: Daily residuals averaged over all years from the stepwise multiple regression.

underpredict. The largest residuals coincide with extreme snowfall events.

Further subsetting of the data for regression modelling was performed, but the results presented above represent the best results obtained. Sub-dividing the snow year into six segments derived from the residual patterns allowed better predictions for some time periods, but the results were not a significant improvement on those already discussed. Grouping the SNOTEL sites by how well the individual station time series correlated with each other produced five regional groupings where the correlation between stations was generally over 0.8. These stations were then aggregated and analyzed using the regression procedures, but again, the results were not much better than the analysis already discussed. This grouping, however, gave much improved results when using the neural networks (see below).

In conclusion, the suspected relationship between atmospheric circulation and snowfall is shown by the stepwise regression analysis. While the variance explained is relatively small for climatological research, there is a definite synoptic control on snowfall over the Colorado Plateau region. Unfortunately, the control determined by the linear methodology is not as good as one would hope for. With only 33.6% explained variance, the implication is that other factors are more important than the atmospheric circulation, or that these linear techniques do not provide an adequate model for explaining circulation-snowfall relationships in this complex mountain terrain.

5.6 Neural Network Analysis

Chaotic and complex behavior in the atmosphere does not lend itself to linear-analytical techniques; hence, non-linear methodologies may have greater utility for establishing relationships in complex systems. Neural networks provide new approaches to problem solving that utilize non-linear mathematics and are applicable to a wide variety of problems. Acting much like a 'black box', able to establish a relationship between data sets and predict new values based on the input data, the neural net provides an alternative methodology for investigating data relationships. With the marginal results stemming from the linear methodology described above, using a neural network approach becomes a logical next step for modeling the relationship between atmospheric circulation and snowfall.

The net used in this study is the shareware program NevProp, version 1.0 developed by researchers at the University of Nevada Center for Biomedical Modeling Research. It is a general purpose back-propagation program that allows the user to interact with the program parameters (i.e., options to force gradient descent, epoch size, etc.) and is available via anonymous ftp (see Appendix I). NevProp allows the user "to run massive data sets and automatically monitor the training process for over-fitting (i.e., to improve external generalization)" (from the NevProp document). NevProp is easy to use and is an excellent program for first-time users of neural networks. It allows unlimited numbers of input patterns, input, hidden, and output layers combined with arbitrary connections between various layers' units. Choices are given for random seed methods for the initial weights and various methods of error propagation and non-linear functions are possible. Moreover, NevProp has been compiled on a variety of platforms including Unix (SunOS Release 4.1, HP-UX [HP 9000/750 and 9000/385], Convex OS [Convex C-220], and Cray OS [Cray-YMP2]), Macintosh OS (with and without co-processor), and DOS.

The analysis described here groups the individual stations (see Figure 5.7 later) into the five regions determined by the inter-station correlation analysis described earlier. These groups match quite closely the individual drainage basins in the study area. The first region is the high-elevation area nearest the continental divide that lies in the upper reaches of the Colorado River and the Yampa-White river drainage basins. The second group includes lower-elevation stations in the Upper Colorado and Yampa-White drainages along with sites in the Gunnison and San Juan drainages. These two groups incorporate the greatest amount of water contribution to the overall Colorado flow (the linear regression predictions for these regions gave approximately 30% explained variance). The third region lies along the Wasatch mountains, the fourth in the Upper Green river drainage and the last lies along the Uinta mountains. The correlation level between stations in each group always exceeded 0.80. Daily snowfall amounts for each group were derived by averaging the available data for each site in the group. On no days were more than two stations missing from any group. Thus, the data were aggregated by region and smoothed by time (5-days).

These aggregated data were placed into a net as the output data value to be predicted from the same smoothed pressure-related variables used in the linear-regression analysis (the three PCA components, their squares and cross products). The best net structure found for the groups was a 9-12-1 configuration with the nine pressure values going to the nine input layer nodes. These are connected to a hidden layer with twelve units that connect to the single output unit--the snowfall. The net was trained with five years of data and the net then used to predict the final three years of the data record. The training years incorporated some extreme climatic conditions with the inclusion of an El Nino/Southern Oscillation (ENSO) event. The test years also included an ENSO event. The predictions from the test-data time series are correlated with the observed data giving a Pearson's correlation coefficient that can be compared to the correlation coefficients derived in the linear analyses. The neural net gives improved correlation coefficients for each of the groups and for the region overall; r^2 values range from 0.24 to 0.54 with a mean of 0.44 (Table 5.1 and Figure 5.7).

Table 5.1: Comparison of r^2 values for the regional groups

Group :	*1*	*2*	*3*	*4*	*5*
Regression *5-day smoothed*	0.29	0.33	0.45	0.42	0.29
Neural Net *5-day smoothed*	0.47	0.50	0.51	0.24	0.46
Neural Net *5-day lagged*	0.64	0.77	0.77	0.68	0.62

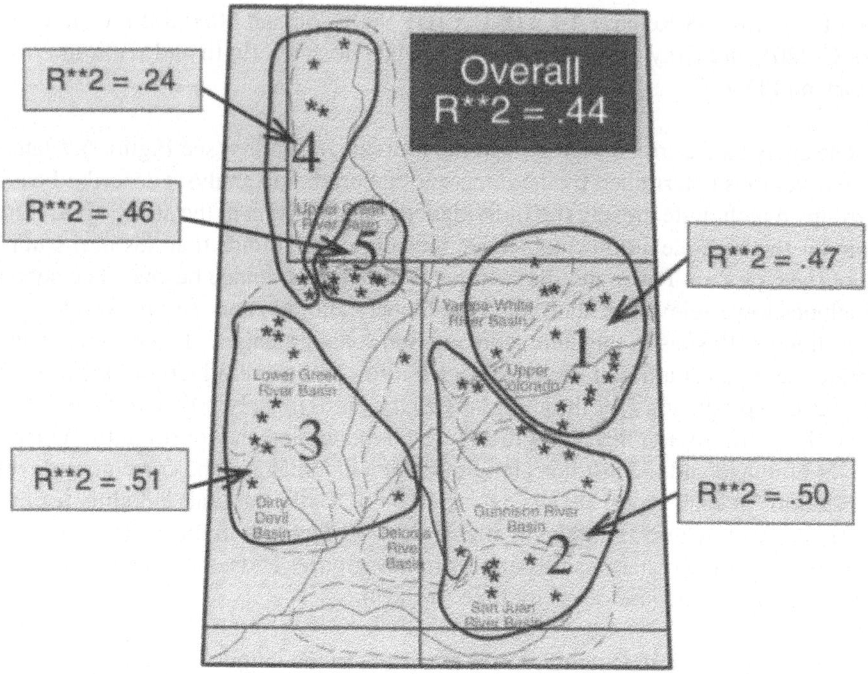

Figure 5.7: r^2 values averaged by grouping highly correlated SNOTEL sites together. Results are from the 5-day smoothed time series.

While the results are an improvement over the linear regressions, they still do not give the level of explained variance that one might desire. Further experimentation revealed that while the optimum results in the polynomial analyses are produced by smoothing the data, the neural net actually gives its better results when applied to the raw daily values. The optimum results, in fact, are produced when using a lagged time series of circulation PCA scores to predict snowfall on any given day. The lagged circulation data provide information on the evolution of the synoptic system through time; by contrast, in the regression analysis, this information could only be captured by the 5-day smoothing of the component scores.

The pressure data were run through a series of lagged data formats to determine the number of lagged days of PCA component scores that gave the best performance in the net. These tests were performed on the Group 1 data--the most poorly predicted region in the regression analysis. The general pattern of the daily PCA component scores and their lagged counterparts are fed into the input layer, which is connected to a hidden layer with more units than the number of inputs (various combinations were tried and the best are reported here). These are connected to a single output layer. For example, the net for a 2-day lagged data set included the PCA component scores for the day N (3 scores), day N-1 (3), and day N-2 (3). These nine input nodes are connected to 12 hidden layer units which are connected to a single output unit (the snowfall). Additionally, the raw data are scaled such that the values range between 0.2 and 0.8 to avoid the asymptotic regions of the sigmoid function in the error backpropagation. The results of these nets are shown in Table 5.2; the 4-day lag (day N, day N-1, day N-2, day N-3, and day N-4) gives the best predictions of the Group 1 snowfall. The 4-day lag allows five days of information to enter the net which encompasses the synoptic time scale discussed earlier.

Table 5.2: Lagged-day Net Results Comparison--Group 1

# Days Lagged	Days	Net Configuration	Pearsons r^2
1	(N--N - 1)	6-9-1	0.533
2	(N--N - 2)	9-12-1	0.591
3	(N--N - 3)	12-16-1	0.635
4	(N--N - 4)	15-22-1	0.643
5	(N--N - 5)	18-27-1	0.656
6	(N--N - 6)	21-32-1	0.655

The results for Group 1 with a 4-day lag give an r^2 of 0.64. The stepwise-regression results for the same stations gives an r^2 of 0.29. In spite of some differences in the way the data were manipulated prior to the analyses, the improvement of the neural net over the linear methods is rather remarkable. The other groups also demonstrate similar excellent results (Table 5.1 and Figure 5.8). Averaging all groups together gives an r^2 of 0.70 compared to the 0.34 from the best regression analysis. This two-fold increase in explained variance demonstrates the improved predictive power of non-linear versus linear methodologies in this particular application. Moreover, this comparison must be considered more significant because the regression results gave only a prediction for a 5-day smoothed value, whereas the neural net predicts raw daily snowfall.

The daily snowfall predictions from the different group nets closely resemble the observed snowfall amounts. The neural net is able to capture the phase and magnitude of the snowfall events over the Plateau very well; day-to-day variations as large as 0.4 inches of snow-water equivalence are well modelled. Figure 5.9 shows the three snow-year predictions for Group 1. Because the net (like the regression methodology) tends to predict values with a somewhat smaller range than the input data, it is expected that the net will again underpredict extreme events. The neural net matches the 1986-87 Group 1 snow year extremely well. In 1987-88, however, several larger events in the early part of the year are underpredicted which causes the overall yearly accumulation to

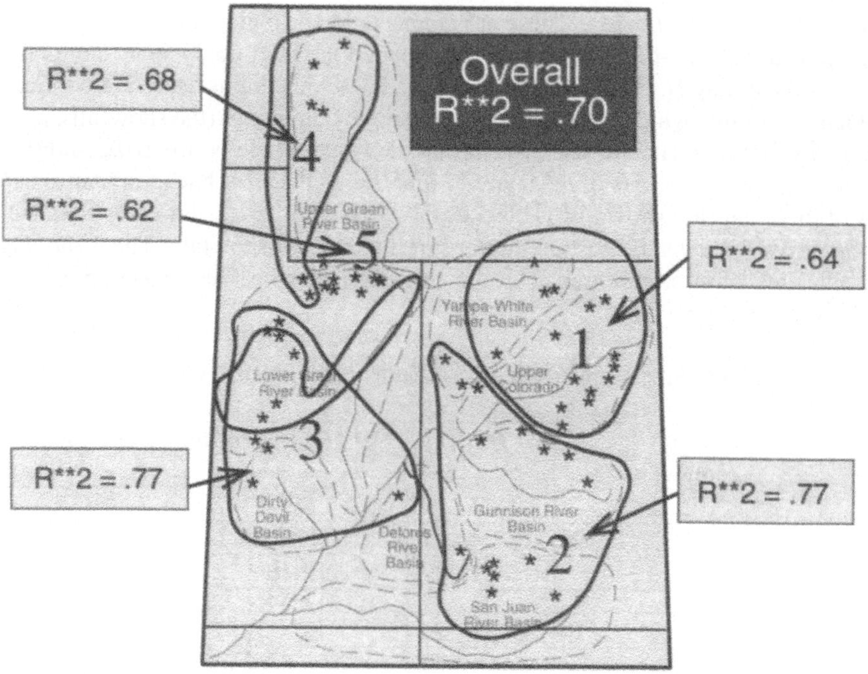

Figure 5.8: r^2 values for correlation-based groups. Results are from the 5-day lagged time series.

be less than observed (Figure 5.10). On the other hand, note that the slope of the observed and predicted snow amount is consistent--the major difference between the graphs is the step function where the extreme event is missed. This is more adequately displayed in the Group 2 results. Figures 5.11 combines the daily predicted/observed snowfall with the seasonal accumulation curves. Three early-season large snowfall events in 1986-87 are underpredicted, each producing an error in the total accumulation that results in a 2.3 inch deficit in that year's predicted snowfall. Although the 1987-88 snow year is well represented by Group 2, the 1988-89 snow year again includes a couple of extreme events that the net fails to adequately capture. Figures 5.12 through 5.14 show the results from the remaining groups and demonstrate the excellent agreement between observed and predicted daily snowfall.

The net predicts a regional value for a climate variable (group average snowfall) from a large-scale parameter (atmospheric circulation) with a high degree of accuracy. Snowfall on the Colorado Plateau is determined by a variety of meteorological parameters such as humidity, temperature, uplift and divergence. While these are all, to

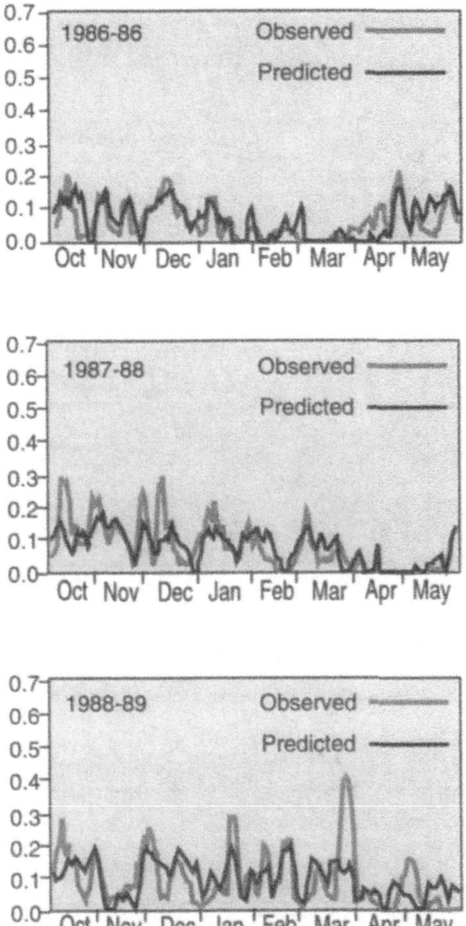

Figure 5.9: Group 1 daily snowfall--observed vs. net predictions.

some degree, a function of the large-scale circulation, none of this information is input directly into the net. Considering that the input data merely show the 5-day history of variance in synoptic-scale 700 mb circulation, the power of the neural networks to predict daily snowfall from atmospheric circulation is exceptional.

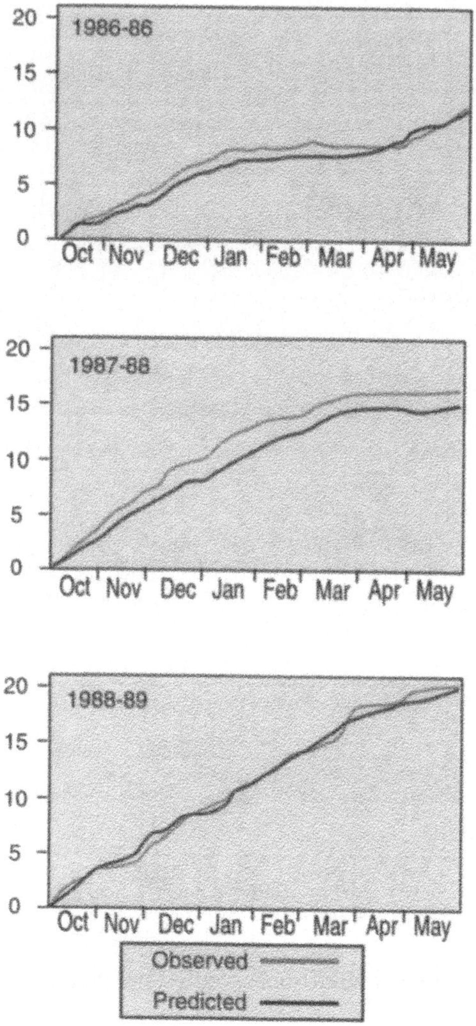

Figure 5.10: Group 1 total accumulation--observed vs. net predictions.

5.7 Conclusions

This study has demonstrated that snowfall over the Colorado Plateau is largely determined by synoptic-scale atmospheric circulation. Both methodologies discussed here show that snowfall is closely tied to circulation. The stepwise regression techniques provide models for linear relationships in the data and predict regional snowfall at the 34% accuracy level. However, in such complex terrain as the Rocky Mountains,

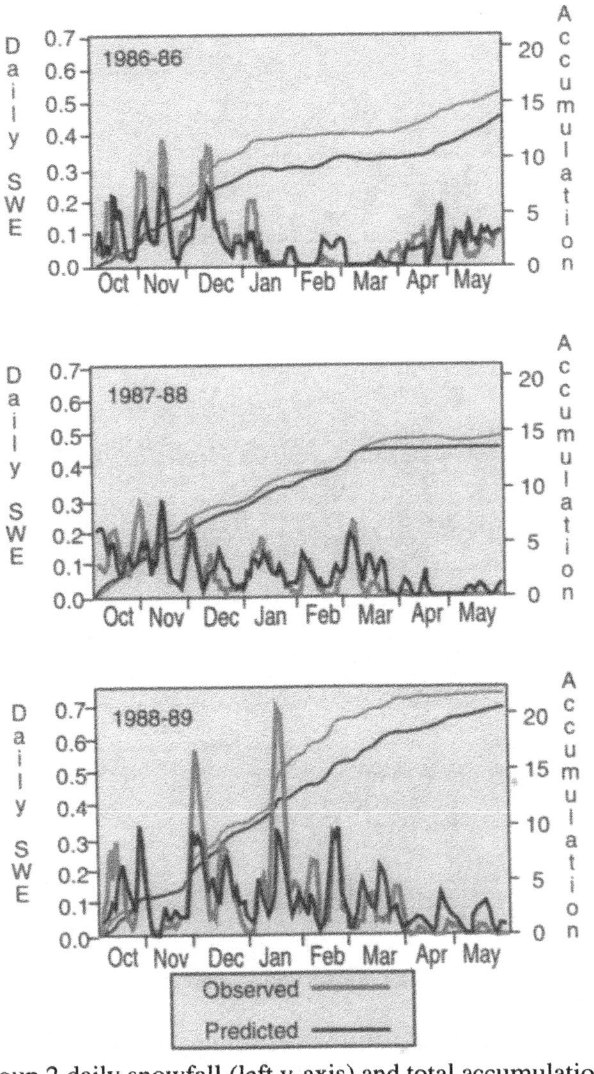

Figure 5.11: Group 2 daily snowfall (left y-axis) and total accumulation (right y-axis)--observed vs. net predictions.

the neural nets predict raw daily snowfall with 70% accuracy--two times better than the linear methods with the same data set. The neural nets are also much more effective at matching the day-to-day snowfall variation. These improved predictions show that snowfall is indeed a strong function of the atmospheric circulation, and that this function contains a significant non-linear component.

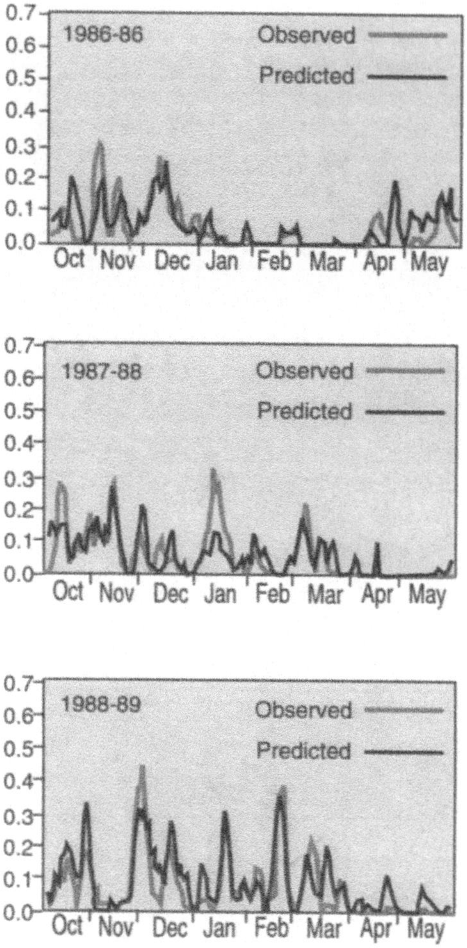

Figure 5.12: Group 3 daily snowfall--observed vs. net predictions.

Not only are the neural nets better at predicting daily snowfall, they provide excellent tools to assess the seasonal snowfall accumulation. Despite the fact that extreme events were sometimes underpredicted, the yearly snowfall accumulations predicted from the circulation inputs are very close to observed. This may provide a means to predict water resource availability from regional snowpack without relying on actual snowfall measurements (however, much more work needs to be done before this claim can be proven).

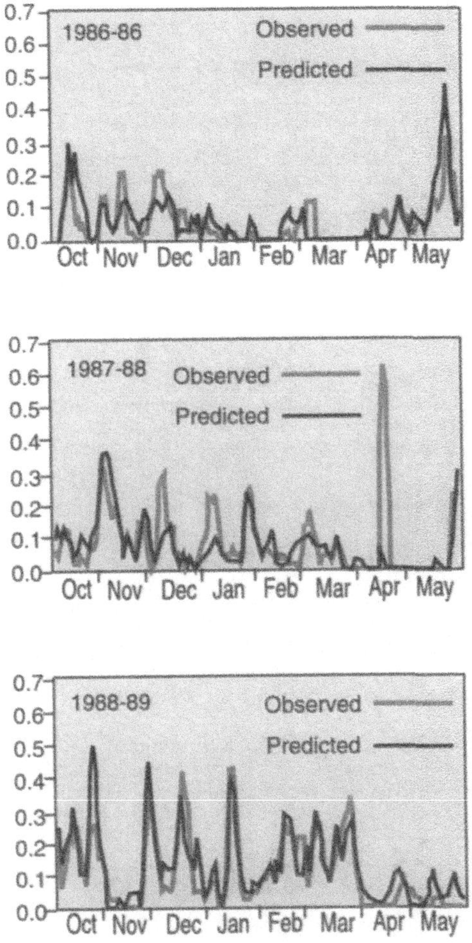

Figure 5.13: Group 4 daily snowfall--observed vs. net predictions.

Finally, the methodologies discussed here all work toward the same goal--that of translation across scales from large-scale atmospheric features to local/regional meteorological or hydrologic variables. Precipitation is, perhaps, one of the more difficult meteorologic parameters to predict, especially in the complex terrain investigated here. However, the neural nets perform excellently despite the difficulties involved.

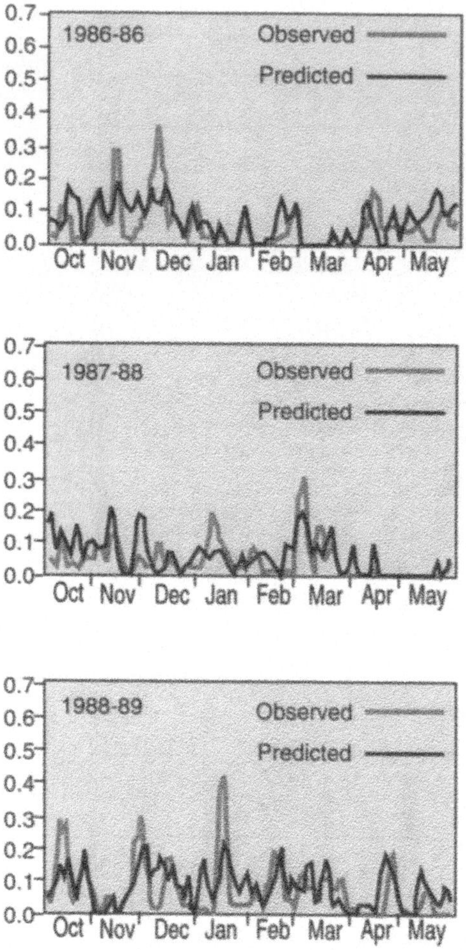

Figure 5.14: Group 5 daily snowfall--observed vs. net predictions.

Acknowledgements

This work was supported in part under NASA Grant NAGW-2686 to E.J. Barron, The Pennsylvania State University.

References

Chagnon, D., McKee, T. B., and Doesken, N. J. (1990) "Hydroclimatic Variability in the Rocky Mountain Region", Climatology Report No. 90-3, Atmospheric Science Paper No. 475, Department of Atmospheric Sciences, Colorado State University, Fort Collins, Colorado.

Cressman, G. P. (1959) "An Operational Objective Analysis System" Monthly Weather Review, 87, 367-374.

Hewitson, B. H. and Crane, R. G. (1992a) "Regional-Scale Prediction From the GISS GCM", Paleogeography, Paleoclimatology, Paleoecology (Global and Planetary Change Section), 97, 249-267.

Hewitson, B. H. and Crane, R. G., (1992b) "Large-Scale Controls on Local Precipitation in Tropical Mexico" Geophysical Research Letters, 19:18, 1835-1838.

Jenne, R. L. (1975) "Data Sets for Meteorological Research, NCAR-TN/1A-111, National Center for Atmospheric Research, Boulder, Colorado.

David McGinnis, The Department of Geography and the Earth System Science Center, The Pennsylvania State University, University Park, PA 16802, U.S.A.

Chapter Six

NEURAL COMPUTING AND THE AIDS PANDEMIC: THE CASE OF OHIO

Peter Gould

6.0 The AIDS Pandemic, circa, 1993

Unless effective vaccines or a cure for the ravaging effects of the human immunodeficiency viruses (HIVs) arrive before the end of the century, possibilities that appear remote at the time of writing, the world faces one of the most devastating pandemics in its history[1]. The World Health Organization (WHO), a normally cautious international body, has predicted 40 million people infected by the year 2000, upgrading prior estimates of 5-10 million made in the middle and late eighties. Although these smaller estimates appeared exaggerated at the time, the new 40 million prediction may itself be an underestimate, although reasonably reliable numbers to check these figures will become increasingly difficult to obtain. In most of the Third World, under-reporting is chronic, and in some regions of Africa and Asia officially reported figures may be as much an order of magnitude too low. Even in North America and Europe, both under-reporting and late-reporting have constituted a problem for monitoring national epidemics; while in countries like France and Britain, HIV infection is still not a legally reportable disease, such reporting to a central authority being left entirely to an attending physician's discretion.

Given the facts that we have no vaccine or cure, and drugs whose abilities to prolong life are highly controversial, we have only two effective and humane ways to combat the mortal effects of the pandemic--education and planning. Education is the *only* thing that can prevent transmission, with strategies ranging from extolling celibacy before marriage, to the free and public distribution of latex condoms and IV needles. The former strategy appears to be ineffective: Uganda, a largely Catholic country

[1]There are obvious difficulties in trying to estimate global figures of people infected, those who recovered, and those who died from pandemics more than 100 years ago, but recent and carefully revised estimates of those dying from influenza during 1918-19 indicate 30 million (Patterson and Pyle, 1991). WHO's estimate for AIDS already exceeds this figure markedly, and I see no reason to suppose that the ultimate total will not be several times this. Little seems to be halting transmission of the virus in most parts of the world today.

B. C. Hewitson and R. G. Crane (eds.), Neural Nets: Applications in Geography, 101–119.

receiving personal exhortations from the Pope in Kampala, has one of the highest rates of infection in the world, with 2.0 million people seropositive by 1993. Latex condoms normally form reliable physical barriers, but frequently are not used on cultural or aesthetic grounds even when they are freely available. In Jamaica, for example, university students, the educated elite of the Caribbean, regard the free distribution of condoms in the dormitories as a joke. In many countries around the world condom use is low because young people are characterized by the 'immortality syndrome'; the danger is seldom perceived as an immediate personal risk. Strategies providing the provision of clean needles to IV drug users have met with greater success, but frequently generate strong opposition, even under conditions of extreme risk (Brown, 1993). By 1990, Holland, with government supported needle programs, had rates of eight percent among IV users; Italy, which simply imprisoned such users without treatment, had 70 percent.

Educating young people in the early years of sexual experimentation requires breaking through the self-confident, but often deadly 'immortality syndrome' with materials that health educators call 'cues to action,' materials that metaphorically grab people by their intellectual necks to make them see HIV transmission as a significant possibility for them. Too often, the HIV is seen as something vaguely 'out there,' far away, and without personal immediacy. Map sequences, showing the often rapid diffusion of the HIV or conversions to AIDS, constitute acts of graphical rhetoric with some power to persuade young people to careful reflection, especially when they incorporate spatial forecasts in animated form for television (Gould, DiBiase and Kabel, 1990). The production of such 'cue-to-action' materials for education requires that we predict the next maps in such sequences (Gould, 1991).

Health planning is required on ethical and humane grounds. Given high rates of present infection, and a median time of ten years between first infection and conversion to opportunistic cancers, pneumonias, and other ultimate and characteristic effects of HIV, humane planning means the provision of hospital beds, wards, hospices, outpatient facilities, special housing, and so on. But *all* planning requires decisions about spatial allocation, taking into account current and forecasted surpluses and deficits, and the degrees of accessibility to such facilities by those who need them, including family and friends visiting such places to comfort the dying (Massam, 1975, 1980, 1993). Algorithms for spatial allocation are well known to geographers, and need not be reviewed here (Rushton, 1973). However, all of them require as inputs for locational analysis predictions of not simply *when*, but *where*. In brief, to use them we need to predict the next maps (Gould, 1993b; Gould *et al.*, 1991). Unfortunately, traditional epidemiological models, focusing exclusively on the time dimension, are scientifically useless, producing banal numbers from differential equations in obvious ways from simple mechanistic couplings: if you hand out clean needles, a temporal model will predict that fewer IV drug users will share and become infected. What other conclusion is possible[2]?

Obviously education and planning are not unconnected. By forecasting the next maps we can see where rapid increases might occur, and so plan for particularly intensive educational efforts and the inauguration of programs of intervention (the distribution of condoms and clean needles). Such map sequences and predictions should also make us thoughtful about the underlying socio-economic conditions that characterize such places, and so inject a note of urgency into the larger field of socio-economic planning as well (Wallace, 1990; Wallace and Fullilove, 1991).

Thus, both the provision of educational materials, and the planning of extensions to current health delivery systems, require spatiotemporal forecasts: forecasts not just of when but where. Several methodologies are available to search for the spatiotemporal information in the (x,y,t) cube, a data base that simply constitutes a pile of maps showing changes over time (t) at specific places (x,y). Generally the approaches form variations on transformation (Gould, 1992), expansion (Foster, 1991), or spatial filtering themes (Foster and Gorr, 1986; Gould and Kabel, 1991). The first attempts to transform the traditional map from geographical space to more appropriate, multidimensional spaces (AIDS spaces, for example), where complex processes of diffusion may take simpler forms. This approach is common in mathematics, where category theory provides a general framework for the transformation of algebraic structures. The latter two approaches allow functions relating AIDS to predictive variables to alter or adapt to local or regional information contained in spatial series (map distributions), valuable geographic information resolutely ignored by epidemiologists playing with their differential equations exclusively in the one-dimensional sandbox called the time domain. Note that all these approaches to geographic forecasting are nonlinear, either by design (the incorporation of quadratic and cubic terms in the initial descriptions), or from the actual investigation of concrete cases (for example, nonlinear functions from parametric tracking with spatial adaptive filtering).

Spatiotemporal forecasting or 'regional forecasting' (Chisholm, Frey, Haggett, 1971) constitutes a general methodological problem cutting across numerous specific fields and subject matters in both human and physical geography--as this volume testifies. It is a difficult problem, tackled only relatively recently with any sophistication, and geographers and other 'spatial scientists' are not so far advanced that they cannot undertake vigorous explorations of the many *hodos* (paths) that lie etymologically

[2]Such a critical view of traditional epidemiological modeling undertaken exclusively in the time domain may appear unduly harsh. For an explication see *The Slow Plague: A Geography of AIDS Pandemic* (Gould, 1993a), particularly Chapter 12: Time but no space: the failure of a paradigm.

concealed in 'methodology' itself[3]. In essence, the problem is how to search for coherent structure or information in an (x,y,t) cube of numbers (or a four-dimensional x,y,z,t hypercube for meteorologists), such that *modest* extrapolations may be made under the usual and often reasonable assumption that what has been 'going on' to date will continue for awhile. I have italicized 'modest' to remind us that all forecasting is an endeavor by finite mortals, whose techniques are extracted painfully and bit by bit from the gods. In the context of AIDS, we would be delighted to see our forecasts fail, perhaps because of advances in medicine we cannot foresee in the mist of the future; for example, the discovery of an effective vaccine or cure that could stop the HIV in its tracks. I note this simply to point to the paradox of scientists desperately wanting their predictions to fail, and even hoping that the predictions themselves will have some effect upon their failure.

6.1 Spatiotemporal Neural Forecasting

As an example of using neural computing to forecast maps in a sequence, we are going to consider the spread of AIDS in Ohio (Gould and Kabel, 1993). Because of its enlightened policy of allowing AIDS researchers early access to county data, and by its careful assignment of cases to counties of residence rather than diagnosis, Ohio formed a valuable 'spatial laboratory' for early spatiotemporal modeling. Nevertheless, we must immediately be aware of three problems that characterize *all* modeling and forecasting efforts. The first is the problem of closure. Obviously Ohio is not floating disconnected in some void, but has numerous connections to surrounding states, to other cities in America, and through these to the 'world system.' Even Iceland and Pacific Islands are not disconnected or quarantined from the rest of the world (Cliff, *et al.*, 1981; Cliff and Haggett, 1985). This is a problem we have to live with, but one that should not be forgotten.

Second, the AIDS epidemic is a 'slow plague,' with its median ten years between first infection and conversion, and initially it was overwhelmingly an infection of the male

[3]In light of chaos and complexity theory, there are a number of extremely important philosophical (mainly epistemological) problems to think through, although they cannot be pursued here. To take but one example, meteorologists for years have tried to find spatiotemporal order in the turbulent flux of daily weather patterns, trying to compute with differentially-coupled equations, and ever-faster computers, those extrapolations from the past to the future that constitute daily weather forecasts. Yet since Lorenz's (1969) truly seminal paper, we know there are relatively short, finite limits (10-12 days) to such predictions, limits not even approached by today's 'machinery.' I do not believe the world's weather patterns will be upset by the flap of a butterfly's wing (I have a deep faith in dampening), but there are clearly limits to our human ability to specify the state of some, always artificially separated, system. There is an interesting difference here between 'processual' approaches, i.e., those searching for information in three- or four-dimensional cubes of numbers to improve forecasts in the future.

homosexual and IV drug communities. In the United States, many were members of a highly mobile generation visiting major epicenters of infection such as New York, San Francisco, Los Angeles, and Miami. There is some reasonable argument that the maps of AIDS in the 1980s are roughly those of HIV infection in the 1970s, implying that despite the spatial mobility of many of those first infected, the classical processes of hierarchical and spatially contagious diffusion were operating, and that most spatial interaction is complementary and reciprocal.

Third, *any* forecasting approach that tries to incorporate the latest information to improve predictions has to be aware of late reporting and under-reporting. In my judgement, under-reporting does not constitute a serious problem in Ohio today, although it may have been a disturbing influence in the early years when diagnostic abilities were weaker, and prurient curiosity higher. Doctors may well have ascribed death to any one of the opportunistic infections, rather than AIDS itself, to save a family 'embarrassment.' However, after 13 years, nearly 400,000 people with AIDS under new definitions, and 1-2 million infected, attitudes have changed. Late reporting is a much more difficult question, since we have evidence that in certain cities it may be chronic. In Washington, DC, for example, 75 percent of the cases reported in June, 1991, were diagnosed not weeks, or even months, but years earlier. It is my personal opinion that Ohio has a much better record than these extreme cases, but late reporting could still constitute a problem for monitoring the epidemic closely, for example on a month by month or quarter by quarter basis. Obviously, all approaches require reliable data, and the neural approach needs up-do-date values for the latest training round. It is no use training on junk whatever the approach.

Ohio, and its particular human geography, forms a valuable spatial laboratory for any forecasting technique concerned with predicting the next map(s) (Figure 6.1). The state is characterized by highly urbanized counties, such as Cuyahoga (Cleveland), Franklin (Columbus), and Hamilton (Cincinnati); moderately dense settlement in rich agricultural areas dominated by large towns of regional importance; and areas of sparsely settled and generally poor rural areas. In a sense, the state is a low-lying demographic plateau punctuated by sharp exponential spikes.

By 1982 (Figure 6.2), conversions to AIDS were reported in Cleveland, with evidence already of spatially contagious diffusion immediately around this epicenter to Akron and Canton, and a jump by hierarchical diffusion to Columbus[4]. Two years later,

[4]There is a nice philosophical and methodological point to be emphasized here. We can only describe what we can see here on the map; or, more precisely, what we are *allowed* to see. The representation of the epidemic as a map sequence gives the *appearance* of spatially contagious and hierarchical diffusion at this scale of allowable observation, a scale enforced by spatial aggregation to the county level to protect individual confidentiality. Note, however, that even if we had access to all the confidential medical files for individuals, we could not do any analytical work with them! While it may, in some

cont .....

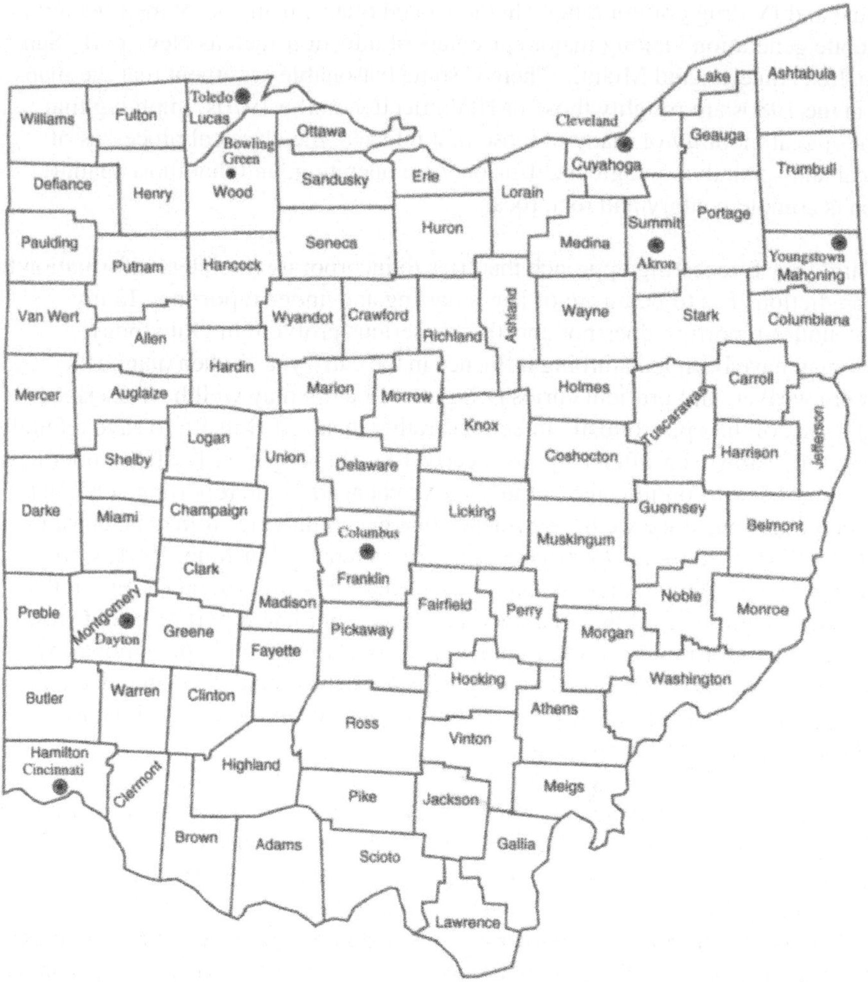

Figure 6.1: The state of Ohio, a mix of highly urbanized counties, constituting exponential demographic 'spikes,' smaller regional centers, and highly rural counties. The 88 counties constitute the geographic laboratory for the neural predictions.

cases, be inferred that Person X transmitted the virus to Person Y, there is no evidence for this in any case involving multiple partners and potential exposures, and there can be no evidence that transmission occurred at a particular time or during some sexual or IV drug encounter. Thus individual files contain essentially no information of analytical or predictive use. Reductionism to the individual level produces analytical banality (Gould, 1985).

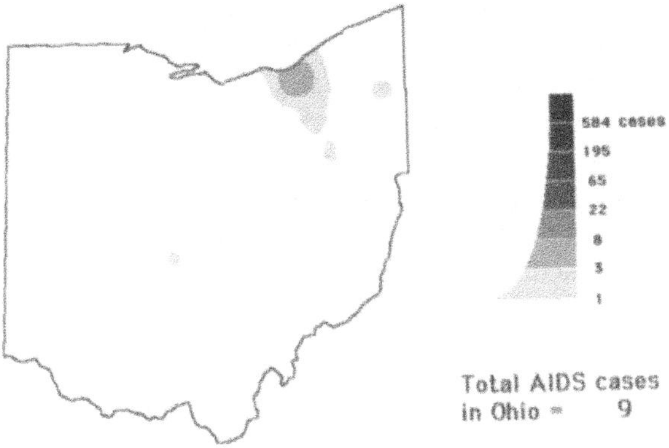

Figure 6.2: The distribution of AIDS cases in Ohio, 1982, based on aggregated county values, but represented as an extrapolated 'AIDS surface.' Notice that the contour interval is *geometric*, each grey-scale change constitutes a tripling of the previous value.

in 1984 (Figure 6.3), the first, barely discernible pattern has strengthened, with clear evidence of seepage from both the Cleveland and Columbus epicenters into the surrounding suburbs and counties, while Cincinnati and Dayton form new centers in the southwest. We can also see the emergence of a northeast-southwest alignment, a diagonal slash from Cleveland to Cincinnati, following the interstate route I-71. Ohio is not a homogenous piece of copper sheet, with the HIV diffusing like the heat from a Bunsen burner, but a space strongly structured by the human presence. Roads, railways and airlines are built to serve human interaction, so it comes as no surprise that the HIV transmission (also a form of human interaction at much smaller geographic distances) is guided at this scale by major transportation alignments.

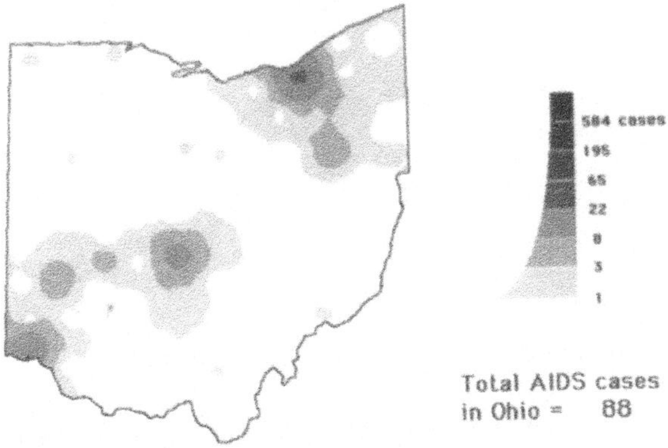

Figure 6.3: The distribution of AIDS cases in Ohio, 1984.

By 1987 (Figure 6.4), the pattern has strengthened, with strong spatially contagious effects produced by the initial and large epicenters 'pumping out' the HIV into the surrounding areas, generally by the intense daily interactions of the journey-to-work (Wallace and Wallace, 1993), as well as continued hierarchical effects to smaller regionally dominant towns serving as local point sources of infection. Three years later, in 1990 (Figure 6.5), there are only a few counties not reporting conversions to AIDS. It is worth emphasizing that the contour intervals are geometric, each change of shading constituting approximately a three-fold increase over the previous gray scale.

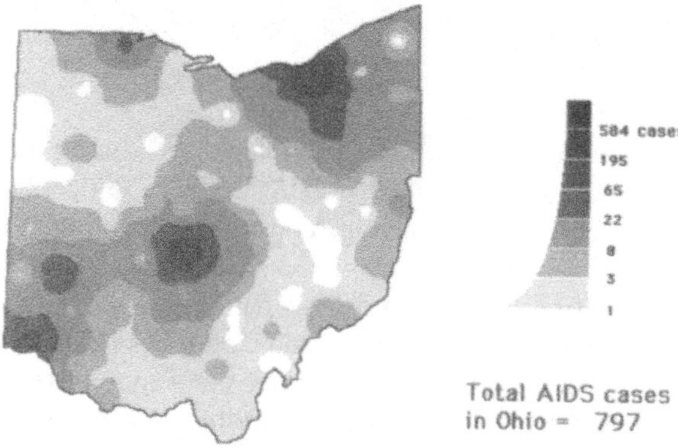

Figure 6.4: The distribution of AIDS cases in Ohio, 1987.

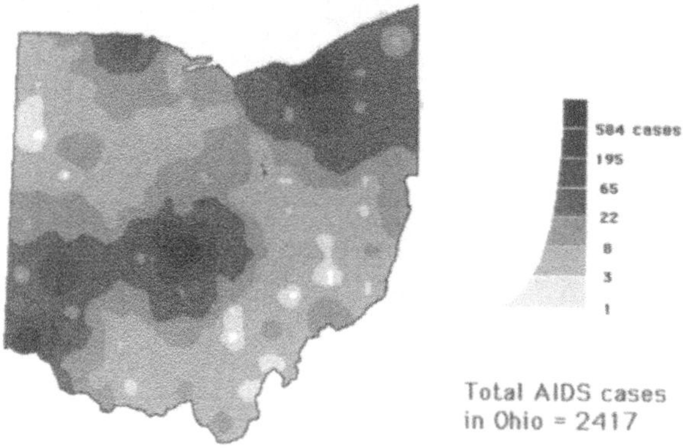

Figure 6.5: The distribution of AIDS cases in Ohio, 1990.

In observing such a map sequence, we can think of each map as a slice in a spatiotemporal (x,y,t) cube (Figure 6.6). Many people, as they run their eyes along or through such a sequence, can make extremely good guesses at what the next few maps will probably look like purely on intuitive grounds. The human brain seems capable of integrating an enormous amount of spatiotemporal information and extrapolating it into the future. Often the analogy is made with a photographic plate developing in the darkroom: future images seem to lie latent in earlier ones. *All* attempts at spatiotemporal forecasting simply try to make well-defined and computable such intuitions about the high degree of information contained in such data cubes.

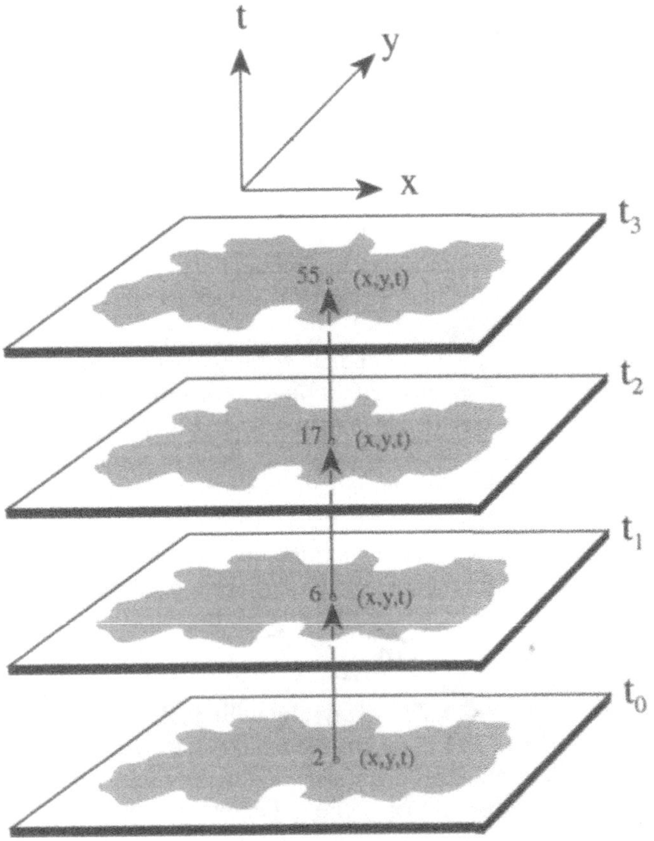

Figure 6.6: A map sequence as a spatiotemporal (x,y,t) cube.

6.2 Neural Forecasting of the AIDS Epidemic

It should be emphasized at the outset that these attempts to forecast were of the nature of an experiment. Few examples of spatiotemporal forecasting exist in the human realm, so these attempts are as much methodological explorations as substantive predictions. The fundamental aim was to predict county values of people with AIDS, or, in other words, to predict the next maps.

Clearly, in using the neural approach, the first task was to choose a set of variables for training the net, while allowing at the same time a freedom to experiment with net configurations, particularly the number of 'gateway' neurons in the first layers. Hewitson (1992) has pointed to these as being possibly related to the major sources (i.e. dimensions) of variation. Most human data of relevance to AIDS prediction at this scale is census based. It is reasonably accurate, readily available, and gathered for pragmatic purposes related to political representation, economic programs, and social description. In the context of the history of AIDS forecasting, these are not the trivial points they seem. Most conventional modelling, constrained strictly to the time domain by the 'differential paradigm,[5]' makes--let me say it bluntly--*impossible* data demands, usually involving the estimation of transmission rates between subgroups that may be changing in unknown ways even as they are being measured (Gould, 1993a). A major problem with census data, apart from well-known undercounting of certain age and gender cohorts of particular ethnic groups, is that the values are only monitored at decade intervals. However, gross geographic patterns do not change particularly rapidly in the United States, and in any case only the 1980 census values were available.

Ten variables were chosen, or computed, from census sources and the records complied by the AIDS surveillance program of Ohio. For some bureaucratic personnel, and those still believing in the twentieth century equivalent of 'phlogiston' (Cavalli-Sforza, 1991), it was of interest to know whether the melanin content of the epidermis had any predictive effect, so total population values for the 88 counties were separately recorded as black and white. Other ethnic categories formed such small proportions that they were ignored. Obviously variables such as the number of homosexuals engaging in multiple sex, or numbers of intravenous drug users sharing needles, could not be obtained, nor, I suggest, would it be desirable to do so on ethical grounds.

In any diffusion process, spatially contagious effects are immediately suspected. These might be recorded in any number of ways, including simply contiguity, contiguity weighted by length or proportion of boundary, secondary contiguity (counties contiguous to immediately contiguous counties) weighted by some proportional constant . . . and so on. All seem equally arbitrary, although one might undertake lengthy, and probably not very informative, empirical investigations unique to the particular set of county elements to see if marginal improvements were possible. As we shall see later, from the results of the experiment, this is unlikely. For this experiment, contiguity effects were recorded simply as the three largest populations of contiguous counties.

---

[5]See again my Chapter 12, Time but no space; the failure of a paradigm, in *The Slow Plague*, pp. 155-167.

Hierarchical diffusion effects should also be suspected in a technically advanced country with modern road, rail and air transport. To capture such effects, the simplest gravity model values for each county were computed directly proportional to the products of the populations, and inversely proportional to distances between county centroids. Again, lengthy calibrations might have been undertaken to see if marginal improvements were possible. In these experiments, the three largest values of potential interaction were recorded for each county.

The first eight variables are purely demographic, geographic, and, it should be emphasized, *static*. The last two making up the set of training variables were simply the number of people with AIDS recorded in the previous two years. These were the only 'dynamic' variables, in other words, variables capable of changing during the experiment. Given a certain faith in the accuracy of reporting, based in turn on a trust in medical practitioners diagnostic abilities and their willingness to uphold the law (AIDS is now legally reportable), this raises the important question of late reporting. If this is a severe problem (as noted above in Washington, DC), then it is no use training on nonsense--for example, a value of the previous year that may be only 25 percent of the true value when corrected two years later. From discussions with those directing the compilation of the county and other relevant AIDS data, I doubt that Ohio's late reporting problem approaches anything like this degree of severity. Nevertheless, county AIDS values were not used beyond 1990 on the grounds that by 1992, when the experiment was conducted, annual values up to 1990 had been corrected and were reasonably reliable.

All training and prediction were conducted on a net with a 4-4-1 feedforward configuration. Increasing the number of first layer or entry nodes showed minimal improvement in the net performance, while decreasing the number degraded the net's learning, and so predictive, abilities. The four gateway neurons in the first layer corresponded to the possible sources of ethno-demographic, contiguous, hierarchical and prior epidemiological effects. The squashing function used in all nodes was a bipolar sigmoid becoming asymptotic at $\pm$ 1.0. The entire data set of training vectors consisted of a 264 x 10 matrix (Figure 6.7) containing three sets of 88 counties and ten variables. County AIDS values for 1986-87-88, and 1987-88-89, were used to match target values for 1988-89-90. From the row set of 264 counties over three years, a sample of 39 (roughly 15 percent) was selected at random as a test data set, while the remaining 225 vectors were used for training the net. In all the experiments, training was stopped after 150 iterations if there was no improvement in the root mean square error (rmse), or after a maximum of 10,000 iterations. We shall consider four separate experiments, each of them conducted on the respectively trained net, and each employing three 'runs' using all 264 vectors, the 225 training vectors, and the 39 test vectors (Table 6.1).

In the first experiment, all the variables were used, and training stopped at 10,000 iterations. All three runs slightly overpredicted the total number of AIDS cases in the state (1.7, 2.3, and 0.3 percent respectively), but all simple linear correlations between

Table 6.1: Results of four training experiments on Ohio AIDS data.

a: All variables used: training halted at 10,000 iterations

	Vectors	Observed	Predicted	RMSE	r
All	264	5623.0	5718.1	3.559	.999
Training	225	3955.0	4044.6	2.812	.999
Test	39	1668.0	1673.5	6.335	.998

b: Black population removed: training to 4,500 iterations

	Vectors	Observed	Predicted	RMSE	r
All	264	5623.0	5740.5	4.074	.998
Training	225	3955.0	4065.7	3.578	.998
Test	39	1668.0	1674.8	6.205	.998

c: Contiguity variables removed: training to 160 iterations

	Vectors	Observed	Predicted	RMSE	r
All	264	5623.0	4308.5	16.478	.974
Training	225	3955.0	2816.1	14.066	.976
Test	39	1668.0	1492.4	26.390	.971

d: Hierarchical variables removed: training halted at 10,000 iterations

	Vectors	Observed	Predicted	RMSE	r
All	264	5623.0	5794.6	4.215	.998
Training	225	3955.0	4083.0	3.010	.999
Test	39	1668.0	1711.6	8.245	.997

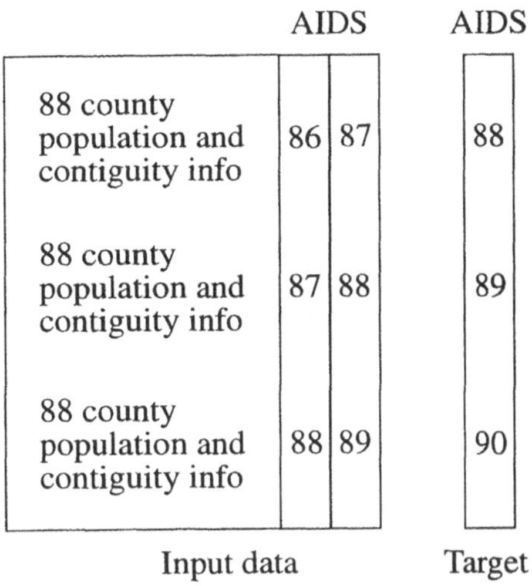

Figure 6.7: The complete data set used for training and testing the neural net.

observed and predicted values were 0.998 or greater. When the individual county errors are sorted by the size of observed AIDS cases (the largest, Cuyahoga 1990, 574 cases, to the 12 rural counties still without any cases reported by 1990), the major urban areas are clearly generating the largest absolute errors (Figure 6.8). Even so, the largest single error (Franklin, 1990, 19.6 cases) is only 2.4 percent overpredicted, and thereafter the error terms decline with a consistent but slight overprediction of the sparsely populated rural counties. Less human interaction of all sorts (travel, commuting, homosexual, IV needle sharing, etc.) may characterize such counties, thereby slowing down the transmission of the HIV (Golub, Gorr and Gould, 1993).

In the second experiment, the black population variable was dropped, a variable increasing in value roughly with the degree of urbanization of a county. Training took 4,500 iterations, resulting in slightly greater overpredictions of total cases, although linear correlations to observed values remained at 0.998. The net seems relatively insensitive to this variable; in other words, the melanin content of the epidermis has little, if anything, to do with the transmission of HIV.

In marked contrast, when the three contiguity variables were dropped, the net trained quickly on 160 iterations, the root mean squared error quadrupled, and the net severely underpredicted the totals (by roughly 10-30 percent). Despite high spatial mobility by the population in general for long distance travel, and in the young male homosexual community in particular, simple 'local' contiguity effects are highly pertinent for understanding and predicting the geographic effects of the epidemic. Who you are

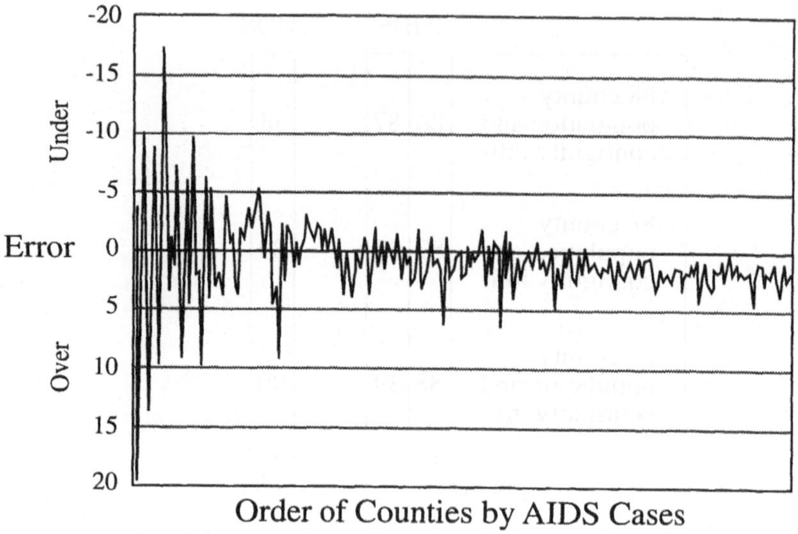

Figure 6.8: The error terms from the trained neural net, sorted in order of the number of observed AIDS cases. Notice that overpredictions are below the zero line.

next to in geographic space matters. In other words, the idea that regional epicenters are 'pumping out' the HIV into the surrounding, essentially commuting, umlands gains strong support (Wallace and Wallace, 1993).

Do hierarchical effects make a similar contribution? The fourth and final experiment dropped the three 'gravity model' variables measuring major interactions of this sort, with training halted at 10,000 iterations. The effect of removing these was relatively small, and the net went back to its former propensity to overpredict slightly. At this scale, (and this *caveat* perhaps should be emphasized), interactions controlled by the urban hierarchy *within the state* seem to have little effect. This does not necessarily mean that such interactions at the national, or even international, scales were not important in producing urban nodes of infection as state and regional epicenters at an earlier time. Simply at this scale of the state, however, they seemed to make little contribution to our predictive abilities.

6.3 A Sensitivity Analysis

By successively removing clusters of variables from the training set, we have been able to assess their contribution to the overall ability of the net to train and predict. But we can also take a second, more refined, approach to evaluating the importance of the input variables. Consider each of the 264 county/year observations as a vector lying in a ten-dimensional 'input space.' If one of the input values in the vector is slightly perturbed, or a slight 'jolt' is given along a particular input dimension, the question is how such a perturbation works its way through the trained net to influence the output value. In other words, how sensitive is the prediction to a small change in input? High sensitivity would imply a relatively high importance to the particular input variable, and vice-versa.

In this analysis, small perturbations were used to 'jolt' each of the ten input variables in each county/year observation. After the ten successive perturbations had worked their way through the already-trained net, the absolute changes in the predictions were recorded. In this way, each observation has an associated vector of ten elements, which we can think of as lying in a 'sensitivity space' (a more detailed description of this methodology is presented in the next chapter). Observations with relatively large elements in the sensitivity vector imply that a county was highly sensitive to such changes in inputs during a particular year. We can consider the lengths of such vectors as sensitivity indices. In a sense, the previous analysis considered the sensitivity of the entire net to train with or without a particular subset of the input variables, while this more refined approach allows us to evaluate the sensitivity of each observation to combined changes in all the variables taken together.

In general, and as we might expect, there is a strong tendency for the large counties to exhibit greater sensitivity; Montgomery County, for example, had values of 9.39, 10.22 and 10.88 in 1988, 1989 and 1990 respectively, while Lickling had only 5.50, 5.53, 5.56 over the same years. There is slight tendency for sensitivity to increase marginally over this time period, due entirely to increasing values in the lagged AIDS case variables themselves. As noted before, these are the only dynamic variables we can observe. When the sensitivity values for each county/year observation are ordered by size of observed AIDS cases (Figure 6.9), there is a sharp drop after the first 20, all of which are large urban counties and regional epicenters. Thereafter, over the remaining county/years, the sensitivity varies little, falling off very slowly from roughly six to five. In geographical terms, the low-lying demographic plateau over most of the state, with its small 'humps' of local centers, appears relatively insensitive. It is the exponential demographic spikes of major cities and regional epicenters that are sensitive to input perturbations.

When the input variables are ordered by their mean sensitivity values (Figure 6.10), we are looking, once again, down the variable columns, rather than across the observation rows. In this way, we can examine, from a rather different perspective, the contribution of each variable to the overall sensitivity index. It is a somewhat different and more refined approach to examining the variables, which in the first analysis were removed in clusters according to ethnic, contiguous, hierarchical and epidemiological

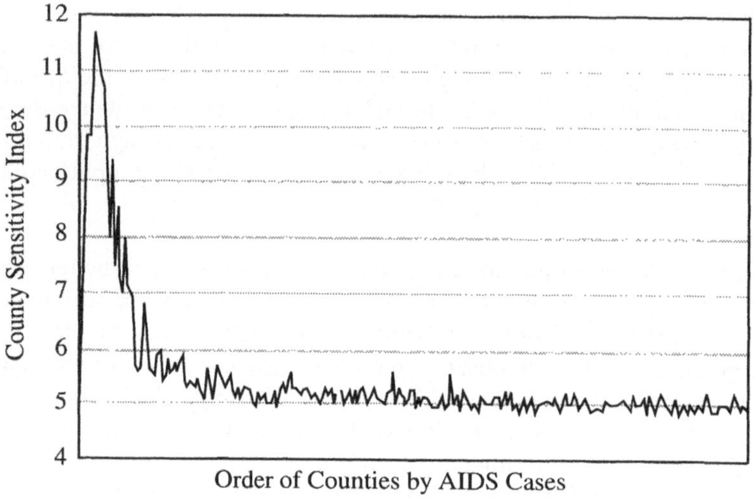

Figure 6.9: Sensitivity indices of the county/year observations ordered by the number of observed AIDS cases.

effects. Notice that the vertical scale is logarithmic: Variable 10 (AIDS cases two years previously), the single variable most sensitive to perturbation, is roughly an order of magnitude more sensitive than Variable 1 (White Population). Considered singly, these dominate the sensitivity analysis: the remainder are again approximately an order of magnitude smaller.

6.4 Neural Spatiotemporal Forecasting: Qualified Conclusions

Once again, it should be emphasized that these attempts to predict the next maps of the AIDS epidemic in Ohio were very much in the nature of an experiment. Very high correlations were received between predicted AIDS cases and those observed, but the correlation coefficient is not a particularly reliable measure of correspondence. For example, *consistent* over- or under-prediction will still result in a high degree of correlation, as in the case of removing the three contiguity vectors, when a 30.51 percent under-prediction occurred, while the correlation remained at 0.97. The measure should probably be discarded in future analysis. Ethnicity appeared to have little, if any, effect: removing the Black Population (Variable 2), raised the RMSE by only 0.51 based on training with all observations. Similarly, hierarchical effects at this geographical scale appear marginally important, possibly because the Ohio 'system' has been artificially clipped out of the national urban structure and its links severed to major epicenters on the east and west coasts. It would be interesting to repeat these experiments at a

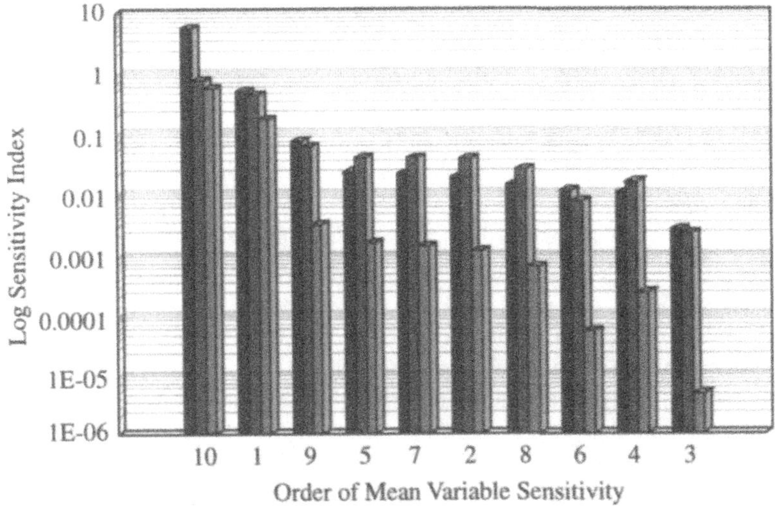

Figure 6.10: Input variables ordered by their mean sensitivity values, standard deviations and variances (in order of columns). The vertical scale is logarithmic: Variable 10 (AIDS cases two years previously) is approximately an order of magnitude more sensitive than Variable 1 (white population).

national, even international level, allowing more far-reaching connections to enter the analysis. In a different methodological mode, that of simple simulation, these international linkages have already been examined for the diffusion of both influenza (Rvachev and Longini, 1985; Longini, Fine and Thacker, 1986) and AIDS (Fahault and Valleron, 1990).

The more detailed sensitivity analysis indicated that observations were highly sensitive, in a relative sense, to perturbations in the lagged AIDS -2 variable. In one sense, this is puzzling: why not the temporally closer variable AIDS -1? Is it possible that these two 'dynamic' variables are so closely correlated that the net selected one (AIDS -2) for a heavy weighting, thereby reducing the corresponding weight on the other (AIDS -1)? Such a possibility should warn us that 'unpacking' black boxes may not lead us easily to the actual variables underlying a process. This makes a strong case for providing a net with *orthogonal* variables for training, perhaps the output of a prior principal components analysis (see Chapter Seven, this volume).

References

Brown, P. (1993) "AIDS: High Risk Behind Bars," New Scientist, 1861, 12-14.

Cavalli-Sforza, L. (1991) "Genes, People and Languages, Scientific American, 265, 104-10. Chisholm, M., Frey, A., Haggett, P. (1971) "Regional Forecasting," London, Butterworths.

Cliff, A., Haggett, P., Ord, K., Versey, G. (1981) "Spatial Diffusion: An Historical Geography of Epidemics in an Island Community," Cambridge University Press, Cambridge.

Cliff, A. and Haggett, P. (1985) "The Spread of Measles in Fiji and The Pacific: Spatial Components in the Transmission of Epidemic Waves through Island Communities," Australian National University, Research School of Pacific Studies, Canberra.

Fahault, A. and Valleron, A-J. (1990) "The Role of Air Transport in the Global Spread of HIV Infection," paper available from Unité de Researches Biomathématiques et Biostatistiques, Université de Paris VII.

Foster, S. (1991) "The Expansion Method: Implications for Geographic Research," Professional Geographer, 43, 131-142.

Foster, S. and Gorr, W. (1986) An Adaptive Filter for Estimating Spatially-Varying Parameters: Application to Modeling Police Hours Spent in Response to Calls for Service, Management Science, 32, 878-889.

Golub, A., Gorr, W., Gould, P. (1993) "Spatial Diffusion of the HIV/AIDS Epidemic: Modeling Implications and Case Study of AIDS Incidence in Ohio, Geographical Analysis, 25, 85-100.

Gould, P. (1985) "The Present and Future Being of Geography as a Human Science, Geoforum, 16, 99-107.

Gould P. (1991) "Modeling the AIDS Epidemic for Educational Intervention," in Ulack, R. and Skinner, W. (eds.), AIDS and the Social Sciences: Common Threads, University of Kentucky Press, Lexington, 30-44.

Gould, P. (1992) "Epidémiologie et Maladie," in Bailly, A., Ferras, R., and Pumain, D. (eds.), Encyclopédie de Géographie, Editions Economica, Paris, 949-969.

Gould P. (1993a) "The Slow Plague: A Geography of the AIDS Pandemic," Blackwell, Cambridge, MA and Oxford.

Gould, P. (1993b) "The Search for Spatiotemporal Information: Predicting the Next Maps of AIDS, Sistema Terra, 6.

Gould, P., DiBiase, D., Kabel, J. (1990) "Le SIDA: la Carte Animée Comme Rhétorique Cartographique Appliquée," MappeMonde, 1, 21-26.

Gould, P., Kabel, J., Gorr, W., Golub, A. (1991) "AIDS: Predicting the Next Map, Interfaces, 21, 80-92.

Gould, P. and Kabel, J. (1991) "Le Epidemia de SIDA Desde una Perspectiva Geografica, GeoCritica, 89.

Gould, P. and Kabel, J. (1993) "At Last Count: The Geography of AIDS," The Atlantic, 271, 1, 90-91.
Hewitson, B., (February 1992) Personal Communication.
Longini, I., Fine, P., Thacker, S. (1986) "Predicting the Global Spread of New Infectious Agents," American Journal of Epidemiology, 123, 383-391.
Lorenz, E. (1969) "Approaches to Atmospheric Predictability, Bulletin of the American Meteorological Society, 50, 345-349.
Massam, B. (1975) "Location and Space in Social Administration, Edward Arnold, London.
Massam, B. (1980) "Spatial Search: Applications to Planning Problems in the Public Sector, Pergamon Press, Oxford.
Massam, B. (1993) "The Right Place: Shared Responsibility and the Location of Public Facilities, Longman, London.
Patterson, K. and Pyle, G. (1991) "The Geography and Mortality of the 1918 Influenza Pandemic," Bulletin of the History of Medicine, 65, 4-21.
Rushton, G. (1973) "Computer Programs for Location-Allocation Problems, University of Iowa, Department of Geography, Iowa City.
Rvachev, L. and Longini, I. (1985) "A Mathematical Model for the Global Spread of Influenza," Mathematical Bioscience, 75, 3-22.
Wallace, R. (1990) "Urban Desertification, Public Health and Public Order: Planned Shrinkage, Violent Death, Substance Abuse and AIDS in the Bronx," Social Science and Medicine, 31, 801-813.
Wallace, R. and Fullilove, M. (1991) "AIDS Deaths in the Bronx 1983-1988: Spatio-Temporal Analysis from a Sociogeographic Perspective," Environment and Planning A, 23, 1701-1723.
Wallace, R. and Wallace, D. (1993) "Inner City Disease and the Public Health of the Suburbs: The Sociographic Dispersion of Point-Source Infection," PISCS Inc., New York, 11.

Peter Gould, Department of Geography, The Pennsylvania State University, University Park, PA 16802, U.S.A.

Chapter Seven

PRECIPITATION CONTROLS IN SOUTHERN MEXICO

Bruce C. Hewitson and Robert G. Crane

7.0 The Issue

While climatology as a discipline covers a host of applications in all regions of the globe, perhaps one of the more critical issues of today relates to climatic change in low-latitude Third World countries. The local precipitation in these regions results largely from small-scale convective features embedded in large-scale phenomena such as the monsoon circulation, the Walker Circulation, and the Intertropical Convergence Zone (ITCZ). Thus, precipitation at the local scale is difficult to both understand and predict, moreover it occurs in regions where people are highly vulnerable to the impact of changing climate and current climate variability. In dealing with such issues, one of the more difficult aspects is to translate from the large-scale processes such as the ITCZ (the levels at which climate processes are best understood), to the smaller scale of immediate human impact. Of particular importance is the problem of translating modeled global changes as a consequence of greenhouse warming down to the local and regional scale.

Southern Mexico is a good example of a region where human societies, particularly those dependent on rainfed agricultural systems, may be very vulnerable to local climate changes. The southern region of Mexico is represented by only one or two grid cells in current Global Climate Models (GCMs). Taking a grid cell value from the model output--a value representing a spatial average over a complex terrain, bounded by different oceans on each side--has numerous limitations and demonstrable errors (e.g. IPCC, 1990). In some cases, errors are so large that simulations of present day climate bear little resemblance to reality, and climate change predictions from different models have order-of-magnitude differences, and even differences of opposite sign. These problems are particularly severe with regard to precipitation.

In many parts of Mexico, and in the absence of widespread irrigation, small-scale agriculture is highly dependent on seasonal precipitation. The timing of the onset of the rainy season is especially critical (Kirkby, 1973), as planting takes place in anticipation of the rains. Since this can vary by as much as a month or more from year to year, the interannual variability in the system, or the misinterpretation of the seasonal signals, can

B. C. Hewitson and R. G. Crane (eds.), Neural Nets: Applications in Geography, 121–143.

play havoc with the life of the local farmer. Consequently, an understanding of the large-scale climatic controls on such local features, and especially how these may change in the future, is of significant importance to local farmers as well as to regional planners.

A first priority in addressing these issues is the development of a methodology that allows us to translate from the large-scale circulation (the scale at which some of the GCMs appear to perform well), down to the local scale. Secondly, we need to interpret these cross-scale controls in terms of physical processes of the atmosphere. In this chapter, we consider how the non-linear 'artificial intelligence' of neural nets can provide both significant insight into the physical controls, and demonstrate substantial improvement over traditional linear techniques in translating from the large-scale down to the local level. Part of this work is derived from initial studies by Hewitson and Crane (1992).

7.1 Southern Mexico Precipitation

The region of study is an area in southern Mexico around Chiapas, approximately 16°N and 22°W (Figure 7.1). The precipitation is strongly seasonal, reaching a peak in summer and a minimum in winter. The onset of the summer rains begins quite suddenly around April and lasts until late October or even into November (Figure 7.2). The primary controls on the climate system are described in detail by Monsiño and Garcia (1974) and their work, along with later studies (e.g., Metcalf, 1987; Cavazos and Hastenrath, 1989; Horel *et al.*, 1989), suggest that the seasonal characteristics are to some degree a function of the large-scale circulation. The summer precipitation appears to be related to the northward movement of the ITCZ while the winter rains are due more to the passage of cold fronts. However, the winter rains are relatively less important than the summer rains both by virtue of their far lower quantity and by the fact that summer is the major agricultural season. Cavazos and Hastenrath (1989) also show that some of the year-to-year variability in rainfall amounts may be related to the modulation of the easterly flow by the Southern Oscillation.

While there is agreement that the large-scale circulation does influence the timing, duration, and characteristics of the rainy season, we do not know how much of the local seasonal precipitation is due to large-scale movement, and how much of it is due to local effects. The summer rain occurs through small-scale convective storms which are unstructured in relation to the synoptic scale, and whose behavior is thus problematic in terms of modeling and forecasting. There has been little research on short-term variability and the degree to which this can be accounted for by large-scale circulation. It is to these problems that we bring the advantages of the neural net in order to distinguish between the large-scale controls and local effects, and to interpret local variability in terms of large-scale processes. The neural net is used to quantify the

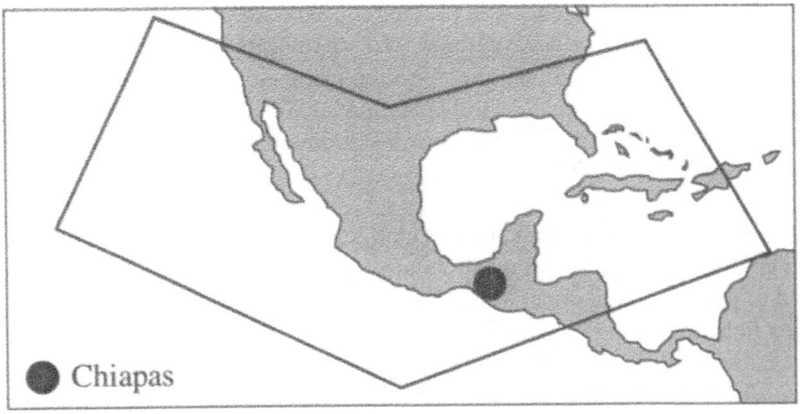

Figure 7.1: Region of study showing window of atmospheric data used and the location of Chiapas.

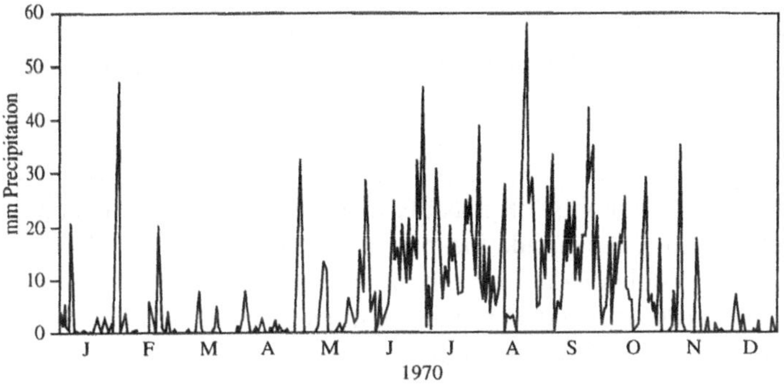

Figure 7.2: Smoothed annual composite of precipitation from Chiapas.

relationship between the large-scale circulation and local precipitation, and to interpret the atmospheric controls represented by this relationship.

7.2 Climate Representation in the Data Set

For this application the data set has to capture two aspects of the climate system: the regional precipitation, and the large-scale atmospheric circulation. Simply representing the regional precipitation raises some difficult problems. For example, station data are representative of the regional precipitation response to the large-scale atmospheric forcing, local topographic and physical features, and the local convective characteristics of the atmosphere. The precipitation record of a particular station reflects a mixture of events both local and regional. We attempt to reduce the local station effects and to represent the regional precipitation using an average of 11 recording stations from around the Chiapas region spanning an almost continuous nine year record from 1970 to 1978. The daily data from each station are averaged to produce a single time series representative of the region.

When considering atmospheric circulation, two aspects need to be characterized: the low-level flow providing both the moisture supply and the surface convergence field conducive to uplift, and the upper atmospheric flow that leads to the upper air divergence field enhancing or suppressing convection. To represent these, a window of gridded sea level pressure (SLP) and 500mb heights at 5pm local time (0000z) are used. The data are derived from the National Meteorological Center (NMC) in the United States and consist of 202 grid points per variable, equally spaced from 9°N to 35°N, and from 70°W to 130°W (Figure 7.1). These span the same time period as the precipitation record. All days with grid point values greater than six standard deviations from the mean were removed, which, along with missing days, left 90 percent of the nine year record.

While the total of 404 gridpoints for the atmospheric data could be input directly to a neural net in order to determine their relationship to the regional precipitation, there is a high degree of correlation between the grid points and hence redundancy in the data set. Furthermore, with such a high degree of correlation between adjacent grid points the neural net, in drawing information from the data, is as likely to choose one as the other of two correlated grid points (or possibly a proportion of both), which would make subsequent interpretation of the circulation-precipitation relationships difficult. To circumvent this, the atmospheric data are initially pre-processed with Principal Components Analysis (PCA) in a data reduction mode to produce a few orthogonal (uncorrelated) 'pseudo' variables for the data.

PCA is a standard technique used in many disciplines and so will not be described in detail here. In essence PCA creates a smaller set of new variables from linear combinations of the originals: a set accounting for most of the original variation. The SLP data and 500mb data are first scaled independently to a +/- 1.0 range to prevent later bias in the PCA output (discussed further below). A matrix is then created of the 404 daily grid point values (SLP and 500mb) from which the correlation matrix is calculated and the eigenvectors and eigenvalues derived. Based on the Rule-N (Overland and Priesendorfer, 1982), which determines the point at which an eigenvector is as likely to be due to random noise as to some real feature of the data, 16 eigenvectors are deemed valid, explaining 94 percent of the variation in the original data set. The 16 components are then rotated and scaled to the Varimax criterion to provide (1) component (factor) loadings, which describe the dependence or contribution of a grid point to a component, and (2) a time series of scores for a component, where the score for a particular day describes how important that component was to the variance on the given day. The loadings may be mapped to show the dependence of a particular place on a component, and they are of importance later for interpreting the neural net. The component scores are derived from the scaled input matrix (scaled in order to prevent bias by the numerically higher valued 500mb data). These 16 orthogonal component scores for each day are used as a multivariate index of circulation in place of the 404 original grid points.

In a final pre-processing step, both the component score time series and the spatially averaged precipitation data are smoothed with a five-day equally weighted moving average filter. The intention of this step is to focus on the atmospheric states or modes conducive to, or inhibiting, precipitation. By removing the high frequency variability, the state of the atmosphere on the synoptic time frame is emphasized. Missing data, and the two days on either side of a data gap, are removed from the filtered data. The smoothed scores are then matched by date to the smoothed precipitation, retaining only those days present in both data sets. These data sets are now used for the neural net analysis focusing on the degree of relationship between the atmosphere and regional precipitation.

7.3 Neural Net Design and Training

A simple neural net is used in this application, as the relationship being investigated is considered only in one direction: that is, the control of atmospheric circulation on local precipitation. A feed-forward net is constructed with the component scores forming the inputs from which the precipitation is to be 'predicted,' while the observed precipitation forms the target output values for the net to train on. The only primary constraint on net configuration is to have a single output node. Otherwise one could place as many nodes in the input and hidden layers as one liked, as well as have any number of hidden layers. Many different net configurations were designed, on a trial-and-error basis, to find the simplest configuration above which addition of nodes and layers made minimal improvement. These preliminary experiments indicated that at least three nodes in the input layer, and five nodes in the hidden layer were required (Figure 7.3), while using more than one hidden layer made only a negligible difference to the results. The output function of each node is a simple bipolar hyperbolic tangent of the weighted summation of the node inputs ranging from -1.0 to +1.0. Each node in the net is given an additional bias input on a weighted link. The bias value on all nodes was set to 0.1 and it provides the possibility for a node to constrain the region of operation on the hyperbolic tangent.

During training, the precipitation values are scaled to +/- 0.8 to fit within the node output range, rather than lying in the extreme asymptotic regions of the hyperbolic tangent function. The data sets are then divided into training data (75 percent random

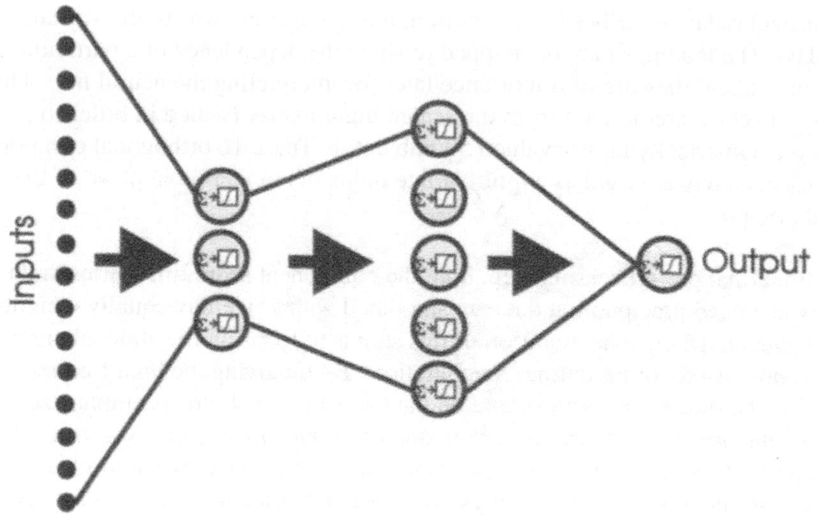

Figure 7.3: Structure of the feed-forward neural net.

selection of the days), and test data (the remaining 25 percent) used to validate the net and prevent overtraining. The procedure is as follows:

a: The net weights are initialized to random values between +/- 0.1.
b: The training data set is presented to the net one day at a time in random order. With each training case the net output error is calculated by comparing the output with the scaled target precipitation value. The output error is backpropogated through the net and the weights updated.
c: After one pass through the training data, the test data are passed through the net with no training taking place, and the root mean square error (RMSE) between the output and the test target precipitation is recorded. At this point, the RMSE is unscaled from the +/- 0.8 range.
d: If the RMSE of the test data is the lowest yet obtained, the current state of the net weights are saved.
e: Steps (b) to (d) are then repeated until no improvement is observed in the test RMSE over 30 training cycles through the training data.

Depending on the random initialization, the trained nets converged to slightly different solutions. Nevertheless, the different solutions produced very similar results. The way in which each of the 16 inputs was incorporated into the net demonstrated that the same relationship was, in essence, being represented in each case. This suggests that in each case the net comes close to the global error minimum, but is trapped in minutely different local minima near the global point, depending on the initial random starting configuration. As a simple means of achieving consensus, 10 different nets (Table 7.1) were trained and their results averaged.

Table 7.1: Correlation between the observed and predicted precipitation record for the nine years of the study period.

Year	No. of Days	Correlation
1970	357	0.85
1971	351	0.74
1972	331	0.75
1973	353	0.83
1974	338	0.67
1975	311	0.78
1976	302	0.73
1977	315	0.68
1978	323	0.76

Over the 9 years the averaged output of the 10 nets gave a correlation of just over 0.8, indicating that 65 percent of the local precipitation can be explained in terms of the large-scale circulation, or conversely 35 percent of the local precipitation is a function of between station variation (Figure 7.4). It is clearly seen from the figure that the net captures the onset and duration of the summer precipitation, which is of critical significance for agricultural production. Note, especially, that the net captures the strong interannual variability of the onset times. In contrast, the net does miss some of the peak values of the daily precipitation, although the phase of the events is well represented. Missing the peaks can be explained as a function of two effects. First, we recall that 35 percent of the total variation is a function of purely local effects, and this will influence some of the under- and over-prediction. Second, one of the advantages of the net is that it is specifically trained to generalize. In the course of training, the net takes a number of specific cases that implicitly represent a sampling of some continuous function, and a generalization of this function is developed. Peak and extreme events represent anomalous situations in the generalized function and are consequently lost.

While the net demonstrates considerable ability in capturing the overall relationship, an interannual variability in the predictive power of the net can be observed if one considers the correlation between predicted and observed values by year (Table 7.1). The table indicates that either the net misses some atmospheric information in some years, or that the degree of influence of the local effects varies in some manner. Nonetheless, the important interannual variability of the seasonal characteristics, such as the onset time and mean values of daily precipitation, are well reflected in each year. For example, the net predictions reflect the marked reduction in late season precipitation in 1977 compared to the more normal peak during this period of the season (see Figure 7.4).

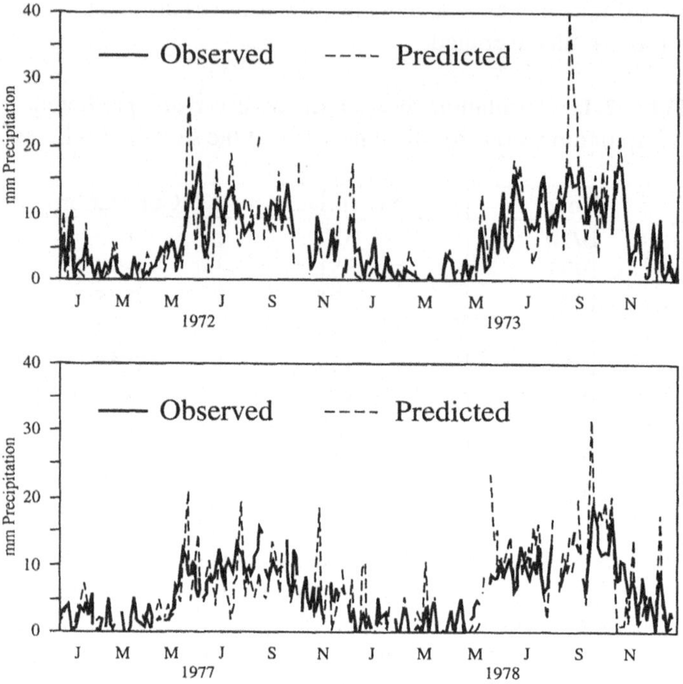

Figure 7.4: Predicted versus observed five-day smoothed daily precipitation for four selected years (Hewitson and Crane, 1992).

The strong control exerted by the large-scale circulation on the local precipitation, and in particular on the timing of onset, suggests that the net could be used to make at least qualitative estimates of the changes that may occur under a doubled CO_2 climate. In attempting to simulate future climates with GCM's there is a large degree of uncertainty attached to a GCMs prediction at any one grid point, especially for precipitation. However, assuming the GCM could get the large-scale circulation correct, and indications are that they do (e.g. Hewitson and Crane, 1992), the neural net should be able to provide an indication of the way the timing of the precipitation onset would change, and how the changes in the large-scale circulation would be reflected in the short term variability within a sub-grid scale region. The assumption implicit here is that the basic relationship of the local effects to the large-scale circulation does not fundamentally change. We assume that the climate change is modeled by shifts along the relationship function (magnitude of precipitation events), and in temporal changes of timing. It is difficult, however, to see how the basic controls such as the ITCZ could change substantially.

7.4 Neural Net Interpretation--Theory

The net, now trained, embodies a complex non-linear transfer function relating the large-scale circulation to local precipitation. Unfortunately, the function is contained in, what is for all practical purposes, a black box. Naturally the weights and node functions are known, and as such the net could be written out as a series of mappings or transformations. However, even in this simple net there are 77 weighted links modifying the flow of information from 16 orthogonal inputs representing 404 grid points in the atmosphere. This degree of complexity precludes a simple interpretation of the function in terms of the atmospheric features extracted by the net for different seasonal characteristics and individual precipitation events. An alternative approach is needed if we are to interpret what the net is telling us about the relationships between circulation and the local precipitation.

Considering the net as a black box, one can approach the issue with the question: under different seasonal and precipitation situations, what are the relative contributions of each input to the output from the net? In effect, and under different climatic conditions, one of the inputs is used by the net to produce its output. Given the complexity of the net, along with the fact that the net manages to perform well through all seasons representing differing climate processes, one would anticipate that under some conditions it will draw more heavily on some inputs than others to produce the predicted precipitation. The inputs it draws on thus reflect the relative importance of various features of the atmosphere to the precipitation.

Another way of considering this is to think in terms of a 16 dimensional input space. Each 16 element case representing the circulation on a given day can be thought of as a 16 dimensional vector locating the state of the atmosphere at some point in input space. If the sensitivity of the output to a small perturbation on any one of the 16 dimensions is measured, it will reflect the significance of that dimension (one orthogonal component of the circulation) to the precipitation predicted by the net *at that location* in input space. A high sensitivity would indicate that the aspect of the circulation represented by that dimension is important, and conversely, a low sensitivity indicates that the atmospheric feature is unimportant. Furthermore, if the sensitivity vectors on all dimensions are measured for a given day's position in input space, then a 16 dimensional 'sensitivity' vector can be constructed, the length of which represents the overall dependence of that day's precipitation on the atmospheric circulation. A long sensitivity vector would indicate that the amount of precipitation is highly dependent on the atmosphere, or that a small change in the atmosphere would produce a proportionally larger change in the precipitation magnitude. For example, under certain highly convective situations (corresponding to certain positions in input space), small perturbations to the atmospheric state would have little effect. Conversly, if the atmospheric state is on the edge of large-scale deep convection, a small pertubation could make a large difference to the precipitation.

Figure 7.5 demonstrates this principle for a simplified, two-dimensional situation (i.e. two inputs). Regarding the two inputs as being two different days, Figure 7.5 shows a plot of the sensitivity of each day to pertubations on each dimension (S_1 and S_2), together with the resultant vector, S_t. In the one case, the situation shows great sensitivity to S_2, and an overall sensitivity of S_t to the atmosphere in general. In the other case, only small sensitivities on either dimension are indicated, resulting in an overall insensitivity to the atmosphere. The overall sensitivity, S_t, is calculated as the net resultant vector of the *absolute* value of all dimension sensitivities S_1 - S_n to avoid a situation of dimension sensitivities of opposing sign leading to a misleadingly small or zero total sensitivity.

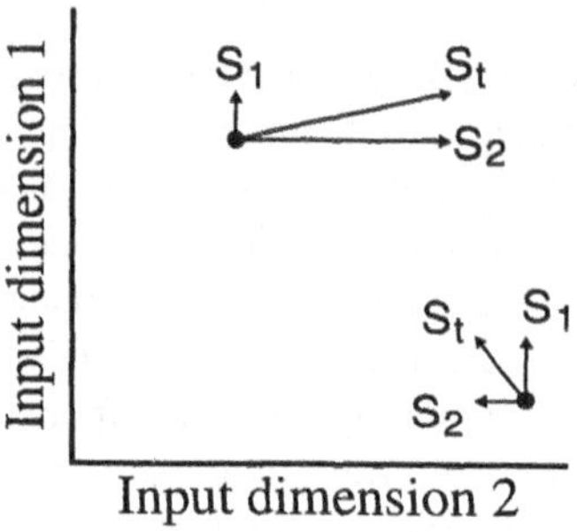

Figure 7.5: Construction of sensitivity vectors.

Note that the sensitivity value represents a sensitivity of change in the precipitation, regardless of the absolute magnitude of the precipitation. Thus from the human perspective in two very different situations, one with trace precipitation and the other with intense rain, similar sensitivities have very different consequences. In the case of trace precipitation, the difference can be between a simple moistening of vegetation surfaces and, say, 5mm of rain--enough to make a difference to vegetation growth; whereas a similar sensitivity on top of 100mm of rain is in essence unimportant. This effect can be accommodated by scaling the sensitivity values by some function of the daily precipitation as described below.

7.5 Neural Net Interpretation--Implementation

The sensitivity procedure is applied to all 10 nets used in the first stage of this study, and the results averaged for the same reasons as before. Each day's score vector in both the training and test data sets is perturbed in turn by +/-0.01, and the effect on the precipitation predicted by the net is recorded--i.e. the sensitivity value. Thus a sensitivity value is calculated for each input for each day, and a time series of the 16 input (component) sensitivities is created. A pertubation of 0.01 is chosen simply to be a very small value with respect to the range of the input values (+/-1.0) in order that the pertubation measures only the sensitivity of the local region of input space located by that day's score vector. In other words, it does not move the score vector into a significantly different region of the input space, where the relationship between circulation and precipitation may be markedly different.

Implicit in applying the sensitivity study in this form is that the behavior of the relationship over the input space is relatively smooth, and that the regions of space represented by the data constitute a reasonably comprehensive sampling over this input space. This is simply assumed at this point; 2981 days represent a comprehensive sample over nine years and, intuitively at least, it is reasonable to assume that the climate processes behave in a smooth and continuous manner.

The sensitivity values are then scaled to account for the relative importance of similar sensitivities in relation to heavy and very light precipitation as discussed earlier. The intention here is to highlight the importance of the sensitivity value in relation to the actual quantity of total precipitation. For example, a given sensitivity value may indicate that the perturbation used to derive the sensitivity gave a *change* in net output of say 5mm of precipitation. This sensitivity when the total precipitation is of the order of 200mm is significantly different to when the total precipitation is of the order of 10mm. In the first case, the sensitivity indicates a percentage change of only 2.5 percent whereas, in the second case it indicates a 50 percent change. Consequently, the sensitivity values are re-represented in a more suitable form by scaling them with a log function of the form:

$$s \,/\, \{log\ (p + 1.0) + 1.0\} \qquad 1$$

where s = sensitivity and p = precipitation.

The divisor is based on the logarithm of the precipitation where the value of 1.0 is added to the precipitation as $log(0.0)$ is undefined. This gives transformed precipitation values that range from 0.0 (the log of 1.0) to some maximum value. However this leads to a division by zero, and for this reason 1.0 is added to the logarithm value, making the scaling factor range from a minimum of 1.0. Figure 7.6 demonstrates the scaling factor as a function of precipitation amount. The length of the scaled vectors is then calculated by taking the length of the vector using the absolute value of all the vector elements as discussed earlier. The time series of the vector lengths is then equivalent to S_t in the two-dimensional example given earlier, and represents the total sensitivity of the predicted precipitation value to the atmosphere on any given day. These values are smoothed with a 30-day running mean filter and the results composited over the nine years. The final smoothed composite is given in Figure 7.7.

Figure 7.7 demonstrates how the 500 mb surface has the dominant control as far as the net is concerned, with the sea level pressure fields providing secondary information. The difference between the two shows a marginal decrease in the relative

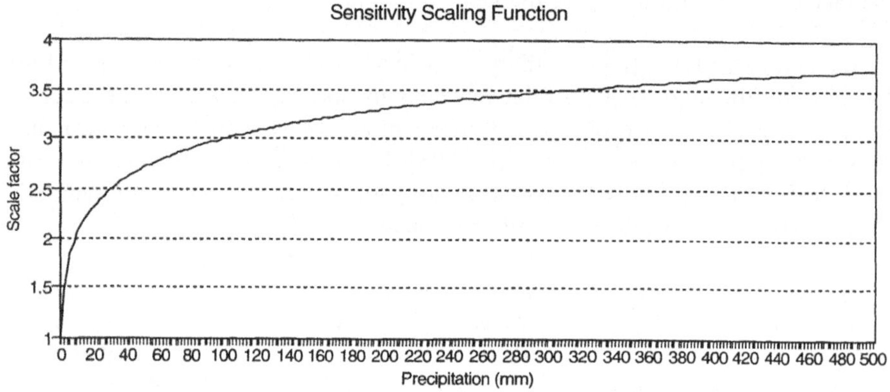

Figure 7.6: Precipitation scaling factor as a function of the precipitation amount.

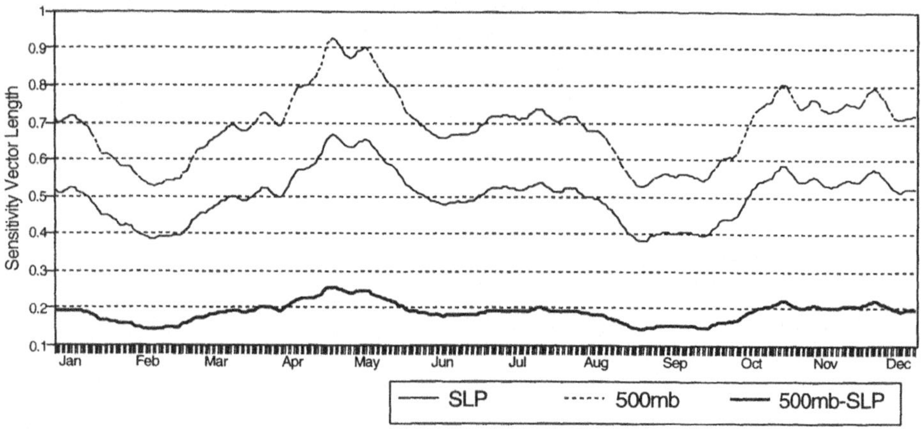

Figure 7.7: Smoothed annual composite sensitivity vector lengths for SLP, 500mb, and 500mb-SLP.

importance of the SLP immediately prior to the summer rains, and during the decline of summer rain at the end of the season. Conversely, it shows a relative increase in SLP importance during the early summer rains. Overall the greatest sensitivity of the summer rains to the atmosphere is during the onset phase, indicative perhaps of the preliminary establishment of the summer rain processes. The immediate pre-season minimum in sensitivity most likely reflects the period of minimum precipitation in the annual cycle.

An interesting phase occurs in the late season precipitation, where there is again a minimum in sensitivity to the atmosphere. Considering that the summer rains are still present, it indicates perhaps that the convective processes are now well-established, and that the occurrence of convection is less sensitive to changes in the circulation.

While this provides insight into the sensitivity of the precipitation to the atmosphere in general, there remains the task of interpreting the net function in terms of specific features in the atmospheric circulation that are of importance to precipitation. To achieve this we return to the PCA results to assist in translating the sensitivity value for each input to a spatial representation. The component loadings from the PCA represent the dependence of each individual grid point on a particular component, or, alternatively, how much of the variance at that grid point can be explained by the component in question. As such they represent a weighting for a component at a grid point, and they are directly comparable to a correlation coefficient. Thus a loading of 0.8 represents a high degree of correlation between that grid point and a component, or that 64 percent of the grid point's variance can be represented by that component.

The loadings can then be used as weights to distribute a sensitivity value for a component across the grid points. Thus a component with a high sensitivity on a given day will be mapped spatially to those grid points that have high loadings on the component in question. This produces a sensitivity map for each day, where the value for each grid point is the summed value of all component loadings multiplied by their associated sensitivity values on that day. As the first 202 loadings of a component represent the SLP grid points and the second 202 represent the 500mb grid points, separate sensitivity maps can be constructed for both fields.

An overall impression of how the sensitivity varied spatially and through the seasons was obtained by animating the map time series in a simple scientific visualization package. From this it was apparent that further interpretation can be performed on three time-frames: inter-seasonal and inter-annual, intra-seasonal, and daily. For this chapter we focus only on the intra-seasonal, and, in particular, the summer rainfall season. The characteristics of the summer season can be established by considering the variation in sensitivity of the precipitation to the atmospheric forcing, together with the characteristics of the precipitation itself. Figure 7.8 shows the composite time series of all years in the data set for the total sensitivity values and the precipitation. Four periods of interest are highlighted in the figure: the precipitation onset phase, the established summer rains, the late season precipitation maximum, and the decay of the summer rains. For each of these periods, composite maps are produced of the SLP, SLP sensitivity, 500mb heights, 500mb sensitivity, and 500mb divergence field calculated from the geostrophic U and V winds. These are shown in Figures 7.9 to 7.13.

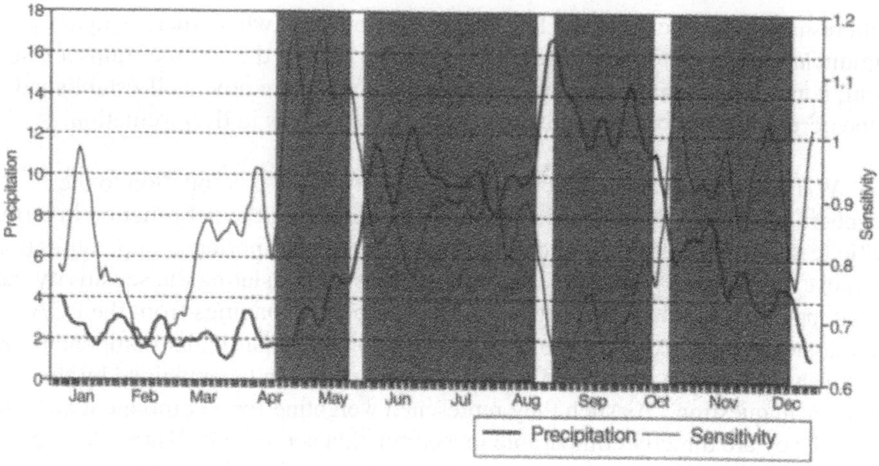

Figure 7.8: Smoothed annual composite total sensitivity vector lengths and smoothed annual composite precipitation. The four periods of investigation are shaded.

7.6 Precipitation Onset

This phase is characterized by a rapid rise in sensitivity to the atmospheric forcing and an onset of the summer rainfall. From Figure 7.7 it is apparent that the 500mb level has a greater rise in importance in comparison to the SLP, and hence should provide the primary indication of which atmospheric features are important. Figure 7.11a indicates that at this stage the 500mb flow has increasing heights toward the equator, with a region of low heights in the north-west related to the southern extent of the westerly trough over the Rockies. The sensitivity map, Figure 7.12a, indicates two strong regions of sensitivity, one west of the Baja peninsula in the region of the westerly trough, and one centered south-west of Chiapas and extending over Chiapas itself, a region shown as a closed upper air high. Initial interpretation of the significance of these regions is problematic until the divergence field is considered (Figure 7.13a). From this it becomes apparent that the net is identifying two regions of upper air divergence separated by a zone of convergence. Bearing in mind that these are composite maps of the daily variability of these features, it would appear that the net is identifying daily variations in the upper air divergence suppressing or enhancing the beginning of deep convection for the summer rains. The region identified to the south-west and over Chiapas is related to the northward advance of the ITCZ over the region, and can be clearly seen on satellite derived cloud images for this part of the season. The region west of the Baja peninsula is related to the southern extreme of the westerly trough, and raises the question as to

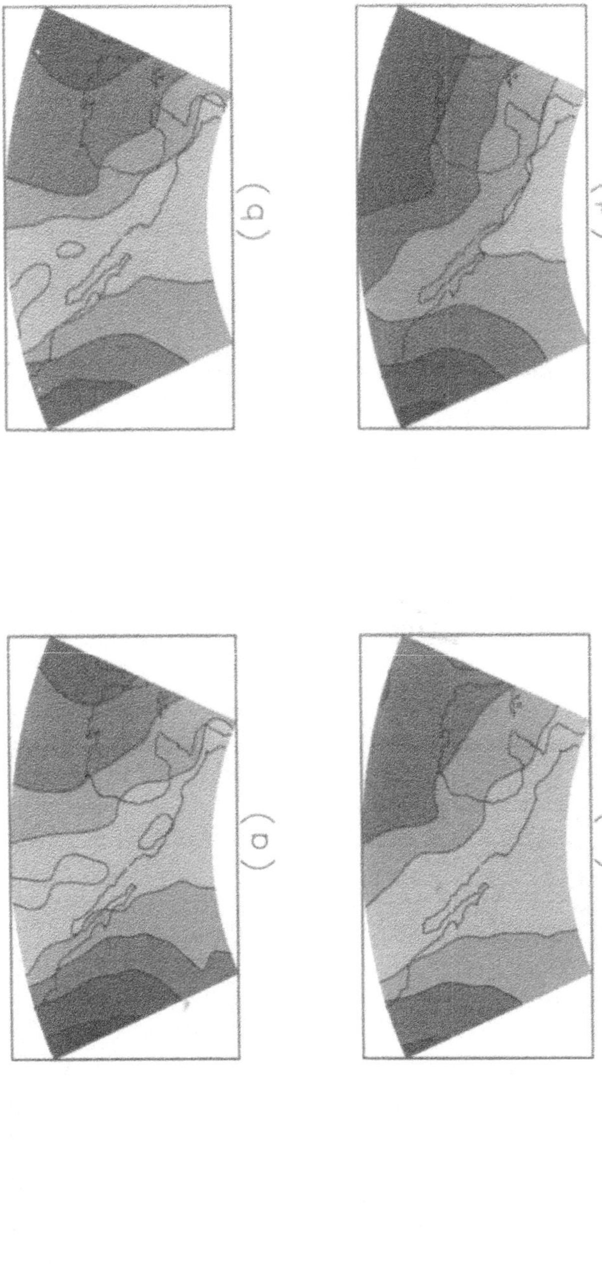

Figure 7.9: Composite SLP maps for the four periods of investigation. (a) onset, (b) early season, (c) late season maximum, (d) tail of season. Dark shading is high pressure.

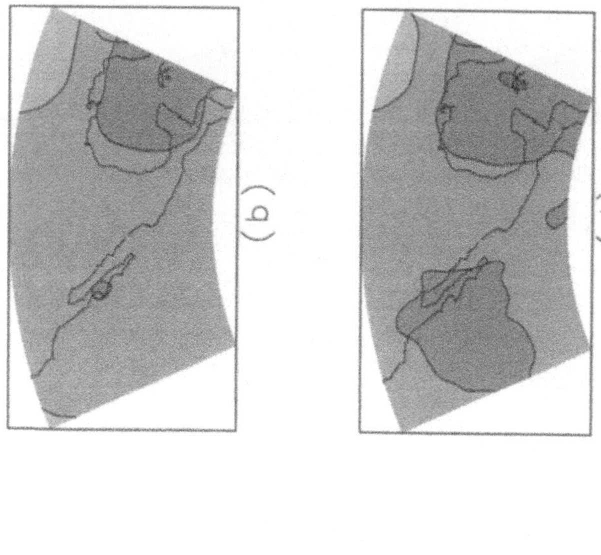

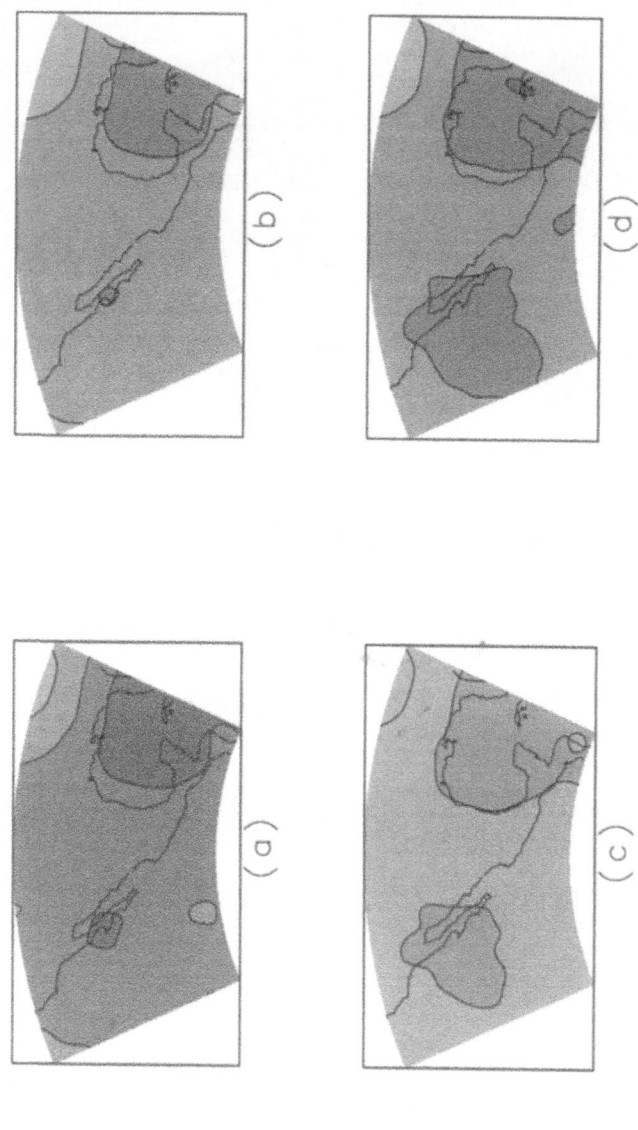

Figure 7.10: Composite SLP sensitivity maps (see text) for the four periods of investigation. (a) onset, (b) early season, (c) late season maximum, (d) tail of season. Dark shading is higher sensitivity.

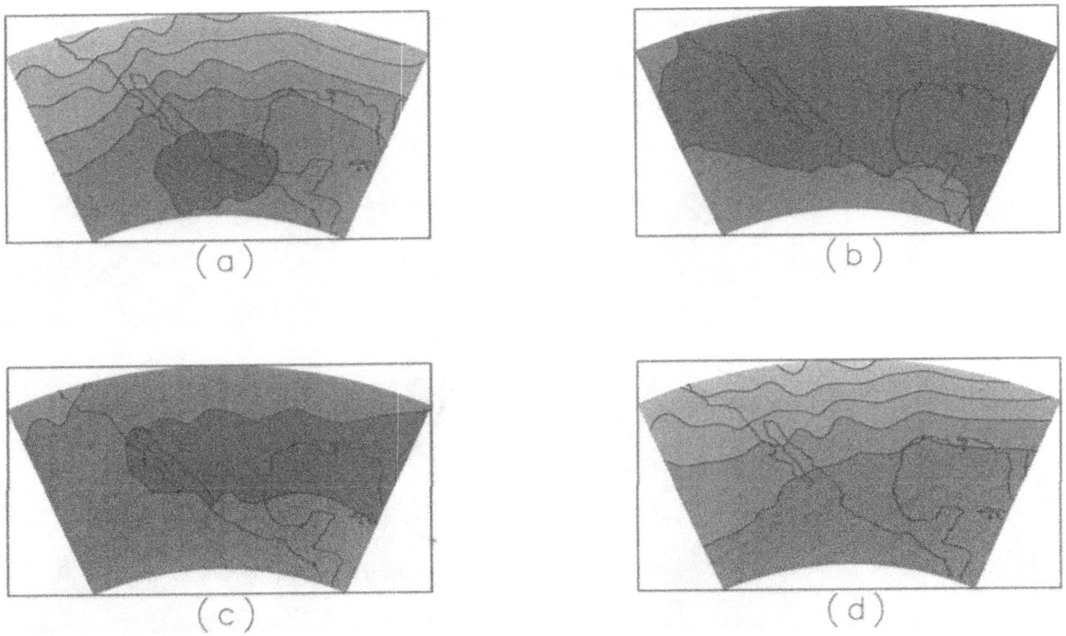

Figure 7.11: Composite 500mb maps for the four periods of investigation. (a) onset, (b) early season, (c) late season maximum, (d) tail of season. Dark shading represents greater 500mb heights.

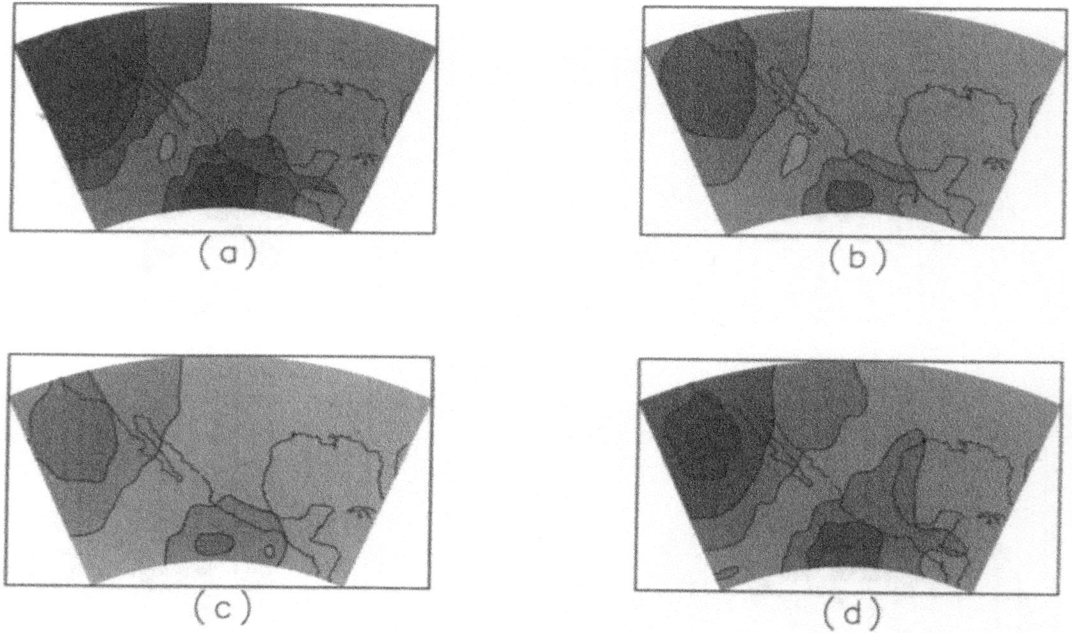

Figure 7.12: Composite 500mb sensitivity maps (see text) for the four periods of investigation. (a) onset, (b) early season, (c) late season maximum, (d) tail of season. Dark shading is higher sensitivity.

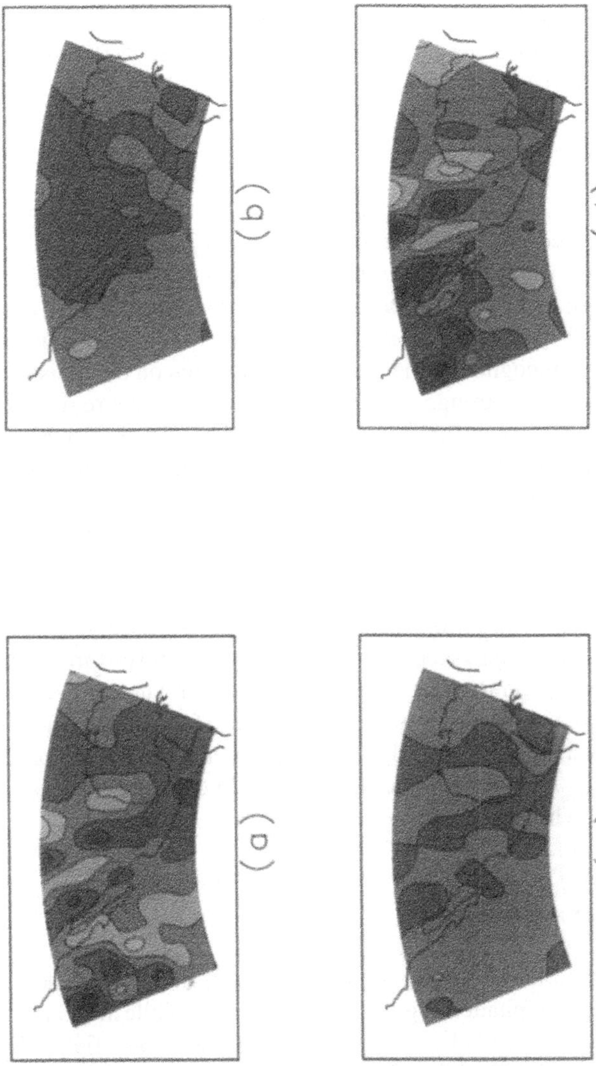

Figure 7.13: Composite 500mb divergence maps for the four periods of investigation. (a) onset, (b) early season, (c) late season maximum, (d) tail of season. Dark shading is positive divergence, light shading is negative divergence (convergence). Calculated from geostrophic u and v wind vectors.

why the net identifies this region with the Chiapas precipitation. We suggest that the relation lies in the connection between the advance of the ITCZ and the northward retreat of the westerly wind belt. In this manner the northward retreat of the westerlies occurs coincident in time with the ITCZ advance, and is required for the ITCZ to establish itself over the region. That the net detects this feature and shows a greater sensitivity to it than to the south-western region is perhaps related to the fact that the westerly trough circulation features are more clearly denoted in the 500mb field than are the weak 500mb features of the ITCZ. It is interesting to note that the region of convergence between the two areas of sensitivity is not selected by the net as having importance.

The SLP fields during this period (Figures 7.9a and 7.10a) demonstrate a significantly lower control over the precipitation. The primary region of sensitivity to the SLP field lies over the Caribbean extending onto the Yucatan peninsula. This region is associated with the southern margin of the Bermuda high pressure system in the region of the easterly surface winds, which transport humid marine air into the Chiapas region. This is interpreted as the net recognizing the controlling features on the moisture supply for the Chiapas precipitation: e.g. stronger easterlies from warm water regions bringing greater moisture compared to a weak Bermuda high shifted eastward. The high sensitivity of the Chiapas precipitation to the atmosphere during this period, especially to the 500mb field, indicates that the atmospheric conditions are on the border of the establishment of the summer rainfall regime; small variations in the atmosphere can change the situation from inhibiting, to being conducive to, precipitation. In conclusion, it would appear that onset of seasonal rains are determined first by the establishment of upper air divergence and deep convection associated with the advance of the ITCZ and retreat of the westerly wind belt, and, second, by an adequate supply of moisture at the surface. The interannual variation in onset time would thus be related to mechanisms determining the retreat of the westerlies, assuming that the ITCZ naturally tries to follow the heat equator. A well-established westerly flow with strong equator to pole height gradients, or alternatively a highly meridional flow regime in the westerlies will delay the onset of the summer rains, while the converse will bring early rains.

7.7 Early Established Summer Rains

During this second period one sees a rapid fall-off of sensitivity to the atmosphere as the summer precipitation regime is established. While the sensitivity drops to both the SLP and 500mb fields, it does so more to the 500mb field. The regions of sensitivity (Figures 7.10b and 7.12b) remain, however, much the same as for the onset period.

In the 500mb field (Figure 7.11b) one finds a significant change from the onset period. The westerly trough has retreated off the map and the height gradient from the equator northward has reversed. The region of divergence in the northwest sensitivity region (Figure 7.13b) has been replaced by a region of convergence. Again this northwest region is problematic in interpretation and suggests some link between this region and the behavior of the ITCZ. The southwest sensitivity region again represents the presence of the ITCZ and the enhancement of convection in the Chiapas region. At the surface, the SLP field still controls the moisture supply. The Bermuda high is essentially unchanged with a slight westward expansion, and the source regions of the mositure would thus be much the same as before.

7.8 Late Summer Precipitation Maximum

The late summer period sees a further increase in precipitation and a fall to very low values in sensitivity to the atmosphere. At this stage it is as if the atmosphere has established itself so firmly in the summer rainfall pattern that it rains regardless of small variations in the atmospheric circulation. The 500mb height and divergence fields (Figures 7.11c and 7.13c) remain essentially unchanged with respect to the regions of sensitivity (Figure 7.12c), but have a slight falling of heights along the northern boundary associated with the re-expansion of the westerly wind field. At the surface the sensitivities to the SLP have also fallen (Figure 7.10c), although they now show a small increase in sensitivity toward the central America region. This may be associated with an expanding region of surface low pressure (Figure 7.9c) modifying the source region for the surface moisture.

7.9 Decay of the Summer Rains

The final stage of the rainfall season shows a return to a situation similar to that of the onset phase. Sensitivities to the atmosphere jump sharply, and especially so with the 500mb field, although not to the same levels as the onset phase. The westerlies reappear along the northern boundary of the map (Figure 7.11d), and the divergence pattern (Figure 7.13d) in the region of sensitivity re-establishes itself. At the surface, however, a more marked difference to the onset phase shows itself. The Bermuda high has now extended westward over the north American continent (Figure 7.9) and results in a more west-east orientated pressure gradient along its margins. This results in a significant change of air flow and moisture source region for Chiapas. This is demonstrated in the stronger sensitivity field (Figure 7.10d). There is also an increased sensitivity to the margin region of an expanded Pacific high, which relates to the retreat of the ITCZ from the Chiapas area and may also influence the moisture supply.

7.10 Conclusions

Perhaps for the climatologist, the most important point to be made is that, as in Chapter Five, the neural net demonstrates a significant ability to capture the cross-scale relationships between the atmospheric forcing and regional daily precipitation. This is of particular note as day-to-day variation in tropical precipitation, due to its small scale convective nature, is conventionally assumed to be related to local processes and to be essentially independent of the large scale circulation. However, in this case, we show that the daily behavior of the local precipitation is, in fact, significantly related to this circulation. Of importance here is the potential for using the circulation predicted by GCM climate change experiments to derive much needed local and regional climate change scenarios in the tropics.

At the same time, it is shown that a neural net need not be a black box, and that significant information may be drawn out with respect to the internal nature of the net. In this case, it is used to demonstrate the changing sensitivity of the local precipitation to the atmospheric forcing, and to identify the nature of some of the atmospheric controls. This now allows for the possibility of investigating climate models to see if the relevant atmospheric controls are captured, and whether their temporal behavior is realistic. At the same time there is the possibility of using doubled atmospheric CO_2 simulations to see how the controls may change under global warming conditions.

Furthermore, the technique offers a means of tackling the particulars about an atmospheric process in a way that circumvents some of the shortcomings of traditional climatological techniques. For example, a traditional means of tackling this problem would be to composite the pressure fields for days of high precipitation, and to interpret the resulting composit atmospheric pressure field as being the primary control on the precipitation. However, such an approach provides no means of identifying the important features within the pressure field, the analysis of which tends to be highly subjective. Similarly, such an approach tends not to focus on the variability in the process or on the factors controlling such variability. These approaches have no means for identifying the changing degree of control the atmosphere may have, which, as has been shown here, may vary substantially.

Finally, only one aspect of the circulation precipitation relationship has been investigated here. The same neural net, with its associated advantages, can also be applied to the interannual variability in precipitation, to examine, for example, the variation in onset time and the large differences in the late season maximum. At the other end of the spectrum, the net may be used to examine the particular day-to-day controls on the regional precipitation, their frequency of occurrence, persistence, etc. It is hoped that such applications will become more widespread in the fields of climatology.

Acknowledgments

This work was supported under DOE Grant DE-FG02-93ER61717.A000 to R.G. Crane, The Pennsylvania State University and NASA Grant NAGW-2686 to E.J. Barron, The Pennsylvania State University.

References

Cavazos, T. and Hastenrath, S. (1990) "Convection and Rainfall Over Mexico and Their Modulation by the Southern Oscillation", Int. J. Climatology, 10, 377-387.

Hewitson, B.C. and Crane, R.C. (1992) "Large-Scale Atmospheric Controls on Local Precipitation in Tropical Mexico", Geophysical Research Letters, 19, 1835-1836.

Horel, J.D., Hahmann, A.N., and Geisler, J.E. (1989) "An Investigation of the Annual Cycle of Convective Activity Over the Tropical Americas", J. Climate, 2, 1388-1403.

IPCC (1990) "Climate Change: The IPCC Scientific Assessment", Houghton, J.T., Jenkins, G.J. and Ephraums, J.J. (Eds), Cambridge University Press, Cambridge, 365 pp.

Kirkby, A.V.T. (1973) "The Use of Land and Water Resources in the Past and Present Valley of Oaxaca, Mexico", Memoirs of the Museum of Anthropology, Ann Arbor, Michigan, 174 pp.

Metcalfe, S.E. (1987) "Historical Data and Climatic Change in Mexico--A Review:, Geog. Journal, 153, 211-222.

Monsiño, P. and Garcia, E. (1974) "The Climate of Mexico, in Climates of North America", World Survey of Climatology, Vol. 11, edited by Bryson, R.A. and Hare, K., 345-390.

Overland, J.E. and Preisendorfer, R.W. (1982) "A Significance Test for Principal Components Applied to Cyclone Meteorology", Monthly Weather Review, 100, 1-4.

B.C. Hewitson, Department of Environmental and Geographic Sciences, University of Cape Town, Private Bag, Rondebosch, 7700, South Africa.

R.G. Crane, The Pennsylvania State University, Department of Geography and the Earth System Science Center, 104 Deike Building, University Park, PA 16802, U.S.A.

Chapter Eight

CLASSIFICATION OF ARCTIC CLOUD AND SEA ICE FEATURES IN MULTI-SPECTRAL SATELLITE DATA

Jeffrey R. Key

8.0 Introduction

The important role that polar processes play in the dynamics of global climate is widely recognized (NESDIS, 1984). The Arctic climate system is characterized by complex interactions between the ocean surface and the atmosphere. For example, the variation of cloud amounts over polar ice sheets, sea ice, and ocean surfaces can have important effects on planetary albedo gradients and on surface energy exchanges (Barry *et al.*, 1984; Shine and Crane, 1984). Cloud cover exerts a major influence over the amount of solar and longwave radiation reaching the surface, and is linked to the sea ice through a series of radiative, dynamical, thermodynamic and hydrological feedback processes (Saltzman and Moritz, 1980). The extent and thickness of sea ice is influenced by the radiation balance, and in turn influences oceanic heat loss and surface albedo. For these reasons the Arctic is considered to be a region where global climate change may be most easily detected. Unfortunately, in situ observations of surface and atmospheric characteristics in the central Arctic are difficult to obtain. Monitoring the polar climate systems using satellite data may be the only viable method of obtaining the information needed to detect climate change.

Unfortunately, the Arctic region provides a unique set of problems for analysis algorithms. Current procedures for automated analyses of satellite visible and infrared radiance data have been developed for low and middle latitudes, but their application to polar regions has been largely unexplored. Those that have been applied often fail because:

1: surface temperatures are commonly as low or lower than cloud-top temperatures,
2. snow-covered surfaces exhibit albedos similar to those of clouds,
3. extremely low surface temperatures and solar illuminations cause satellite radiometers to operate near one limit of their performance range,
4. in winter no visible-wavelength data are available, and,
5. rapid small-scale variations, which in lower latitudes signify changes in cloud cover, occur on the surface as a result of changes in snow and ice distributions so that clear scenes are much more variable here than in lower latitude regions.

B. C. Hewitson and R. G. Crane (eds.), Neural Nets: Applications in Geography, 145–179.

Because of these problems, complex analysis methods specific to particular climate regimes are necessary (WMO, 1987; Rossow, 1989).

Motivated by the apparent limitations of multispectral feature extraction from imagery and the availability of expert system development tools, artificial intelligence (AI) techniques have seen increased use for the analysis of remotely sensed data (e.g., Nicolin and Gabler, 1987; Matsuyama, 1987; Estes *et al.*, 1972; Nandhakumar and Aggarwal, 1985; Campbell and Roelofs, 1984), and have also been employed in geographic information systems (GIS) applications (e.g., Usery and Altheide, 1988; Ripple and Ulshoefer, 1987; Robinson and Frank, 1985; Smith *et al.*, 1987; Jackson and Mason, 1986; Smith, 1984). Due to the limited knowledge of the physical processes in the environment and the inherent noise in many geophysical data, environmental systems often cannot be represented accurately through numeric values describing their physical properties and interactions, but rather are subjected to categorization into broad classes. Most applications of expert systems have sought to apply qualitative knowledge to decision-making; expert systems operate primarily on abstract symbolic structures. In remote sensing applications where pattern recognition, spatial and temporal context, and multivariate analysis are common requirements, coupled numeric/symbolic systems may be useful. This issue has recently been addressed by Kitzmiller and Kowalik (1987), Kowalik (1986), and Borchardt (1986). We show an example of a coupled numeric/symbolic system later in this chapter.

Neural networks also have considerable potential for remote sensing, as suggested by applications to automated pattern recognition (e.g., Ritter *et al.*, 1988). How can neural networks be used to extract features from images? Objects such as clouds, roads, or fields in satellite data are distinguishable by their spectral and spatial characteristics. They all have characteristic spectral signatures; e.g., they reflect a certain amount of light and appear either bright or dark in visible data, and the temperatures at which they radiate may separate them from their backgrounds in thermal data. These types of features have certain sizes and textures which further help identification. When designing a neural network for feature extraction any or all of these characteristics can be used as input. Additionally, a network may itself compute some of these characteristics, or they may be quantified externally. For example, there are mathematical definitions for a variety of texture measures that can be computed outside the network and passed in through one of the input nodes. Alternately, the network architecture may be such that the spectral value of each pixel within a small region of the image is represented by an input node, and the network determines the texture and passes it on to the next level. There are many possibilities, some of which are exemplified here.

In this chapter we investigate applications of neural networks to image analysis. We restrict our area of investigation to the analysis of satellite data for the purpose of extracting information about clouds and the surface in the Arctic (Figure 8.1). First, a

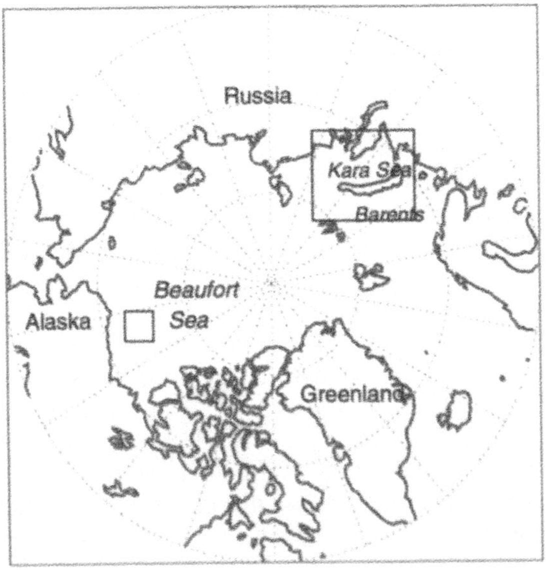

Figure 8.1: Geography of the Arctic. Boxes outline the study areas discussed in this chapter.

feed-forward, back-propagation neural network is used to classify merged visible, thermal, and passive microwave satellite data. Four surface and eight cloud classes are identified. Partial memberships of each pixel to each class are examined for spectral ambiguities. Classification results are compared to manual interpretations and to those determined by a supervised maximum likelihood procedure. Second, a neural network is developed that utilizes local texture to classify clouds based on their morphological characteristics. The cloud-only portion determined by the first investigation is examined in an attempt to label cloud types. Third, clear-sky Landsat images of sea ice are used to extract information about the geometrical characteristics of fractures in the ice ('leads'). Lead characteristics derived through the use of a neural network are compared to those based on a simple expert system. While the applications are to Arctic data, the methodologies are equally applicable to imagery from any region. In fact, with minor modification the same procedures could be applied to a variety of features: land cover mapping, stream network analysis, or land-use patterns, to name just a few.

8.1 Cloud Detection and Classification

Techniques for cloud detection from satellites have been developed for use with visible, near-infrared, and thermal data. Procedures for detecting cloudy pixels are based on threshold methods, radiative transfer models, or statistical classification schemes. The simplest procedures examine the spectral values (e.g., reflectance or temperature) of pixels and label them as either cloudy or clear when compared to some value that has been predetermined to separate clouds from their background. More complex threshold methods use additional cloud or background classes, incorporate texture, or use more than one spectral band. Statistical classifiers or clustering schemes group pixels based on spectral and/or textural characteristics, where the criteria for placing a pixel in a class is most commonly a function of probability or Euclidean distance. To aid in the determination of clear-sky radiances, the spatial and temporal aspects of cloud decks and surfaces can also be utilized. For example, a large local variance in either space or time is usually indicative of cloud cover. Perhaps the most comprehensive procedure for detecting and analyzing clouds in satellite data is that of the International Satellite Cloud Climatology Project (ISCCP) (Rossow *et al.*, 1985). Even with its complexity, however, reliable detection of cloudiness in the polar regions is particularly difficult, and it has been recommended that the study of clouds over polar regions be continued (Key and Barry, 1989; WMO, 1988). It is for this reason that cloud analysis using neural networks has been explored.

8.1.1 Cloud Detection Based on Spectral Features

In this section, we investigate the ability of neural networks to extract cloud and surface information from satellite data. The example given here is intended to illustrate the basic methodology for image classification using neural networks, and is taken from Key *et al.* (1989a). Other procedures are possible, a few of which are reviewed in Section 8.4. The motivation for using neural networks to classify Arctic clouds and surfaces is as follows. First, since cloud and sea ice mapping for climatological studies requires the processing of many images covering large areas (for example, 30 days worth of images for the northern hemisphere), selection of training sites in a supervised scheme or the assignment of spectral clusters to physical classes in an unsupervised approach, can involve an unacceptable amount of time and effort. Second, since a class such as low cloud over ice actually includes a range of cloud thicknesses overlying a range of ice concentrations, considerable spectral variability exists within the class as well as within individual pixels. Therefore, the primary goal of this work is to investigate the ability of a neural network classifier using only a relatively small training set to deal with the considerable within-class variability encountered in our data. Manual and supervised classifications are used to provide benchmarks for comparison of the neural network results, rather than as a test of the merits of these more traditional methods.

8.1.2 Data and Methodology

The data sets used here provide a broad range of spectral information necessary to map clouds and surfaces in polar regions. These data are typical of the types of imagery used for mapping of global cloud, sea surface temperature, and other climatological variables. The Advanced Very High Resolution Radiometer (AVHRR) on board the NOAA-7 polar orbiting satellite measures radiance in five channels encompassing the visible, infrared, and thermal portions of the electromagnetic spectrum (1: 0.58-0.68μm, 2: 0.73-1.0μm, 3: 3.55-3.93μm, 4: 10.3-11.3μm, 5: 11.5-12.5μm) with a nadir resolution of 1.1 km. Global Area Coverage (GAC) imagery is a reduced-resolution product created through on-board satellite processing, with each pixel representing a 3 x 5 km field of view (Schwalb, 1984). Channels 1 and 2 were converted to an approximate spectral albedo; channels 3, 4, and 5 were converted to radiances in milliwatts/(m^2-steradians-cm) then to brightness temperatures (NOAA, 1984; Lauritsen *et al.*, 1979). The typically low water vapor content in the polar atmosphere and the low physical temperatures reduce most atmospheric effects to a point where they may be neglected for the analyses performed here. Channels 1 and 2 were divided by the cosine of the solar zenith angle to adjust for illumination trends in the imagery.

The Nimbus-7 Scanning Multichannel Microwave Radiometer (SMMR) is a conically scanning radiometer that senses emitted microwave radiation in five channels: 6.6, 10.7, 18.0, 21.0, and 37.0 GHz, with two polarizations (horizontal and vertical) per channel. At these frequencies, passive microwave data is relatively unaffected by clouds and provides useful data year-round independent of solar illumination. The 18 and 37 GHz vertical polarization channels have fields of view of 55x41 km and 27x18 km, respectively, and are used here primarily for surface parameterization; the microwave data distinguish clearly between sea ice and open water.

In order to study both clouds and surfaces beneath clouds, it is worthwhile to combine the AVHRR and SMMR channels into a single image set. AVHRR and SMMR data were merged in digital form and mapped to a polar stereographic projection. This projection yields pixels true to scale at 70°N latitude with a five kilometer pixel size. Five kilometer pixels were averaged over 2x2 cells yielding an effective pixel size of ten kilometers square. (Further constraints imposed by the image analysis system reduced the image size to 125x124 pixels.) SMMR data were converted to the five kilometer cells by simple duplication of pixels. Further details are given in Maslanik *et al.* (1989). In this form, color composites can be made consisting of combinations of microwave, visible, and thermal-wavelength channels to highlight different cloud and surface features in the data.

The study area (Figure 8.2) is centered on the Kara and Barents Seas extending north toward the pole and south toward Norway and the Siberian coast. Novaya Zemlya is near the center of the image. Shown are AVHRR channels 1, 3 and 4 for July 1, 1984. Both AVHRR and SMMR imagery were also acquired for July 4, 1984. While covering only a small portion of the Arctic Basin (1250 x 1250 km), it includes representative samples of all surface types found in the Arctic: snow-covered and snow-free land, sea ice of varying concentrations, and open water. The figure illustrates many of the problems involved in mapping polar surfaces and clouds. With the exception of

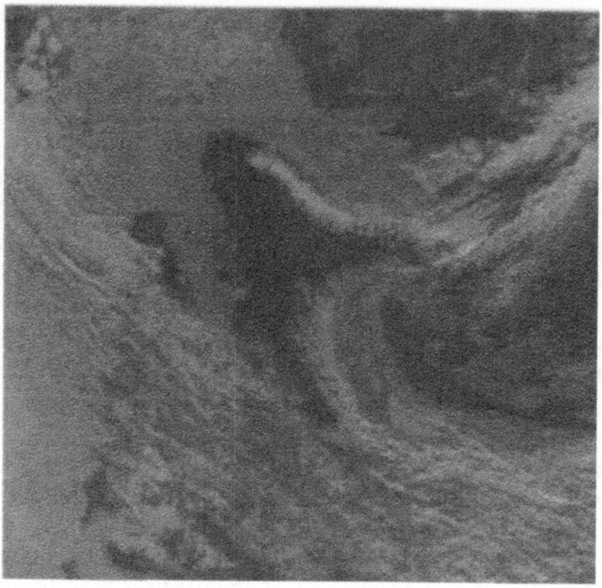

Figure 8.2: A composite of AVHRR channels 1, 3, and 4, showing the study area. Top center is the Kara Sea with the island of Novaya Zemlya (snow/ice-covered) below. The Barents Sea is in the center of the image. Sea ice covers the left side of the image, extending into the Kara Sea. Low clouds over water are in the image center, middle-level clouds are in the lower quarter, and high clouds are at left center and upper right. Snow-free land can also be seen at upper right.

land/water boundaries, edges between classes are typically indistinct, as is the case at the sea ice where the transition from high concentration ice to open water often takes place over a very broad region. For classes that include thin cloud, such as the areas of low cloud over ice and water, cloud opacity varies within the class. The sensors, therefore, record varying proportions of surface and cloud within a single class. Cloud height is also highly variable; heights actually may fall anywhere within the low, medium, high designations.

Four surface classes are of interest in this study: snow-free land, snow-covered land/ice cap, open water, and sea ice (abbreviations used: LAND, SNOW, WATER, ICE). Three broad classes of cloud - low, middle, and high - are defined by temperature as measured in the AVHRR channel 4, and are further categorized by the underlying surface type. Not all surface/cloud level combinations occur in the study image, and those that do not are excluded from the analysis. Eight cloud/surface classes are examined: low cloud over land, water, and ice; middle cloud over water and ice; and high cloud over land, water, and ice (abbreviations used: LCLL, LCLW, LCLI, MCLW, MCLI, HCLL, HCLW, HCLI, respectively). The data are classified by two procedures: a neural network and a maximum likelihood classifier. The maximum likelihood procedure is supervised, using the same training areas as used to train the neural network.

The network uses a feed-forward, back-propagation architecture with a layer of 7 input units representing the AVHRR and SMMR channels, a layer of 10 hidden units, and a layer of 12 output units representing the surface/cloud classes. Part of the network is shown in Figure 8.3. The connections shown will be discussed later. The network is trained on patterns (training areas) for each desired class. After training, the network is presented with the complete data set (the image) and computes a membership value, represented by the activation of the output units.

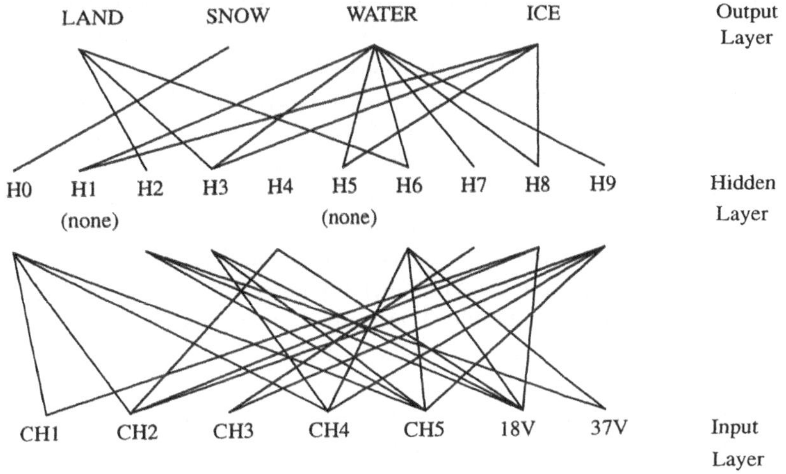

Figure 8.3: Part of the neural network architecture used to classify surface and cloud types in AVHRR data. The input nodes are the spectral values of each AVHRR and SMMR channel employed while the output nodes are the desired classes. Only the surface classes are shown due to the complexity of the network.

Where spatial and spectral boundaries between phenomena are diffuse, hard classifiers which produce mutually exclusive classes seem particularly inappropriate. This issue is discussed further in Key *et al.* (1989b) in relation to the fuzzy c-means algorithm. The neural network approach addresses this problem of often indistinct spectral boundaries between classes by providing as output a numeric value as well as the class symbol for each pixel. This is a membership value for the pixel to each class, and is in the range [0,1], larger values implying greater strength of membership in a particular class. With the sigmoid activation function used in this study, these values are approximately linear throughout the range.

In this example, two sets of training areas (referred to as TA1 and TA2) were selected in the typical supervised manner, with each training area manually delineated in the digital imagery. The training sets were chosen so that the effects of different within- and between-class variability could be tested, with TA1 representing a relatively small sample designed to study the ability of the neural network to address within-class variability not contained in the training statistics (e.g., the variability expected due to changes in ocean and ice temperature, ice concentration, and cloud height and thickness over space and time). TA1 included 1% of the 15,500 pixels (125 lines x 124 pixels) in the test data set. Additional training areas were included in TA2 to expand the variance of the training statistics sufficiently so that a significant portion of the test images would be classified using the maximum likelihood classifier. TA2 included about 9% of available pixels. Class means for each spectral channel were nearly the same in TA1 and TA2 but, with the exception of the LCLI class, standard deviations were twice as large on average in TA2 (mean standard deviation in DN of 1.9 for TA1 versus 3.8 for TA2). As noted earlier, the selection of training areas this large is not practical for climate applications requiring analysis of many images over large areas (thus the impetus to test the neural network using TA1). However, TA2 was needed to address the trade-off between classification accuracy and human interaction using the supervised maximum likelihood approach for the types of data used here. The July 1 and July 4 images were manually interpreted using digital color composites of several AVHRR and SMMR channels. The manual interpretation thus acts as a hard classifier, with classes that consist of a 'best-guess' estimate of class membership based on visual clues. Maximum likelihood (ML) classifications of the July 1 and July 4 images with seven data channels as input were carried out using TA1 and TA2 statistics.

The neural network was trained using the individual pixel values in TA1 as input patterns. To address indistinct spectral boundaries, pixels were assigned to the class with the highest membership value. Pixels with no membership value greater than 0.4 (arbitrary) were tagged as unclassified. In an attempt to similarly relax the restrictions of the maximum likelihood classifier, cut-off limits for pixel position within the n-dimensional Gaussian class-membership curve were varied to a maximum of 99%, and different apriori probabilities were tested.

8.1.3 Results

A manual classification of the July 1 image based on the classes described previously is shown in Figure 8.4, and will be used for comparison to the neural network and maximum likelihood results. The results presented here are largely qualitative; for complete details see Key *et al.* (1989a). Eighteen percent of the image was left unclassified where no dominant class could be determined. As noted above, within-class variance is large, particularly in classes LCLI and LCLW. Coefficients of variation are greatest in the 18 GHz SMMR data and for the AVHRR visible-wavelength channels within these classes, suggesting some confusion between ice and open water.

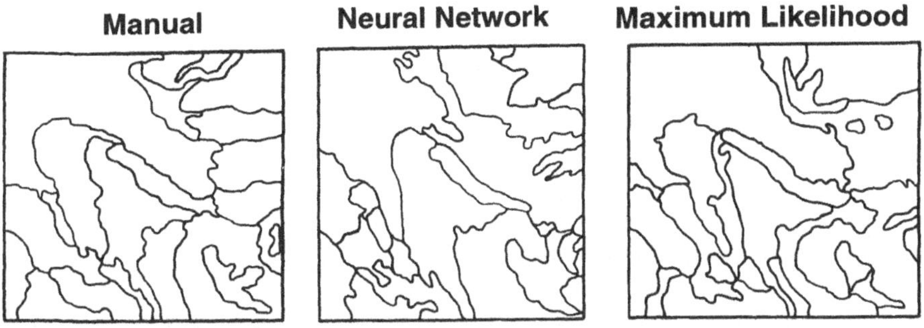

Figure 8.4: Manual, neural network, and maximum likelihood classifications of the data shown in Figure 2. For the neural network classification, a pixel's class is the one in which it exhibited the largest membership value. A pixel is left unclassified if none of its membership values exceeds 0.4.

Table 8.1 shows the total percentages of the July 1 and July 4 images that were actually classified using training areas TA1 and TA2, and the ML and NN (neural network) classifiers. The neural network classification of the July 1 image is shown in Figure 8.4. Some important differences are apparent between the neural network output and the manual classification. The NN results underestimate the amount of low cloud over ice. The NN classification also puts a larger portion of the ice margin area into the WATER rather than ICE class, and tends to assign cloud/surface classes to surface classes, particularly in the case of ICE versus LCLI and WATER versus LCLW. Confusion also exists between cloud height classes. Of the pixels classified in both the NN and manual classifications (i.e., excluding unclassified pixels), overall agreement between classification schemes is 53%. The NN classification of the July 4 image shows similar patterns. As was the case for the July 1 data, nearly the entire image was classified. Differences between the NN results and manual classification for July 4 were greatest between cloud height classes and between low cloud over ice versus clear sky over ice.

Results of the image classification using the maximum-likelihood procedure are shown in Figure 8.4 for July 1 using TA2 statistics. Since the ML classifications using TA1 essentially included only those pixels within and adjacent to the TA1 training areas, these images are not shown. Comparing the manual interpretation with the ML classification shows that the ML classification using the more comprehensive training areas of TA2 effectively captures the basic cloud and surface patterns. However, more

Table 8.1: Percent of images classified by method and training set.

Method	Training Set	Image	% Classified
ML	TA1	JULY 1	2%
ML	TA1	JULY 4	3%
ML	TA2	JULY 1	70%
ML	TA2	JULY 4	53%
NN	TA1	JULY 1	96%
NN	TA1	JULY 4	93%

than half of the manually-interpreted MCLW class is left unclassified by ML. Remaining unclassified pixels are divided among cloud classes and ice/water classes. This supervised ML classification achieved a high agreement of 85%, representing the large training set in TA2. Extension of these TA2 signatures to the July 4 data using ML illustrates the reduction in applicability of training signatures over time compared to the NN classifications, as shown by a general decrease in the percentage of the image that is classified. Given the variability of ice conditions and cloud thicknesses within a single image, it is not surprising that day-to-day variability is enough to reduce the representativeness of the training areas in terms of class mean and covariance. With the exception of class LCLI, standard deviations of training areas in TA1 are considerably less than is the case for TA2 and the manually interpreted classes. TA1 signatures thus include only a small portion of the variance in the desired classes, as indicated by the low percentage of the image actually classified using TA1 statistics.

From these results it can be seen that the neural network classifier is more flexible than the maximum likelihood procedure, and better captures the indistinct spectral and spatial boundaries between classes. However, the relatively low overall accuracy of the classifier implies that the training sets and feature vectors may need to be re-examined.

8.2 Cloud Pattern Analysis Using Texture

Statistical frameworks for describing the morphology of cloud fields as well as the radiative, dynamical, and microphysical processes determining this morphology are needed to better understand climatic forcing (Committee on Global Change, 1988, pg. 117). In this section, the previous analysis is extended in an attempt to determine various cloud types; e.g., stratus, stratocumulus, cirrus, etc., using measures of local texture. A number of studies have included texture in cloud classification schemes, generally in a clustering framework (e.g., Welch *et al.*, 1988, 1989a, b; Garand, 1988; Ebert, 1987, 1988, 1989; Parikh, 1977). Contextual analyses of frontal patterns and cloud shadows are given in Gurney and Townshend (1983), Wang *et al.* (1983), and Swain *et al.* (1981). While cloud detection schemes exist for many data types and geographic locations, the inherently subjective nature of defining cloud types and the algorithmic difficulty of incorporating texture into the analyses are two inhibitory factors in the development of the automated cloud typing methods needed for large scale cloud climatologies.

8.2.1 Methodology

Only the cloudy portions of the images are examined. A neural network is then developed in an attempt to retrieve cloud morphologies. The network architecture is the same as was used before: feed-forward, back-propagation with an input layer representing the AVHRR data (no SMMR data), one hidden layer, and an output layer representing the cloud types. The network is trained on patterns (training areas) for each desired class. A maximum likelihood classifier was again used for comparison.

The method of utilizing texture in this analysis is an attempt to eliminate the problems inherent in computing texture for cells within a fixed grid, that being the mixture of different classes within a single cell. Texture values are assigned to each pixel rather than to a grid cell by moving a 16 x 16 pixel cell across the image shifting two pixels at a time. Each pixel may be assigned as many as 162/2=128 values for each feature. The mean of these values is the value finally assigned to the pixel. While this method does not completely eliminate the problem of mixtures of classes within a cell, it does provide a value which is generally representative of the texture within the neighborhood, although when edges between cloud classes are present, the value will be skewed.

What type of texture measures should be employed? The most widely-used in image analysis are the second order statistics that summarize the probability of the intensity (grey level) values of a pair of pixels. The relative frequencies of grey levels are computed for each pair of pixels in a given positional relationship and are summarized in a grey level difference (GLD) histogram (Weszka *et al.*, 1976). The data are first quantized to 64 grey levels, and the grey level difference, g, is computed for each pair of pixels in the cell over each of four angles: up-down (0°), left-right (90°), upper left-lower right (135°), upper right-lower left (45°). Texture may contain a directional component so that the histogram must be specified as a function of angle as well as distance. A histogram of grey level differences is then constructed for each distance and angle, and used to compute various texture measures. The histograms will be spread over a larger range of g as graininess or streakiness increase.

Textural features initially tested include a variety of second order grey-level statistics, Fourier measures, vector strength, and the Roberts Gradient. These are all defined in Key (1990). A combination of divergence parameter testing and principal components analysis was used for feature selection to eliminate those features that are highly inter-correlated, thereby reducing the training time of the neural network.

Textural features used in the final analysis are angular second moment, entropy (both GLD measures), and vector strength (all in AVHRR channel 1). The mean, maximum, and range of these quantities over the four angles are used in subsequent analyses. In addition to the texture features, spectral values for AVHRR channels 1, 3, and 4 were used.

The angular second moment measures the homogeneity of gray level differences with distance and direction. Angular second moment will be high for decks of stratus and for bands of clouds oriented in the direction of θ.

$$ASM(\theta,d) = \sum_g \left(\frac{h_{\theta,d}(g)}{H_{\theta,d}} \right)^2 \qquad 1$$

where $g=0,1,...,63$, $h_{\theta,d}(g)$ is the number of grey level differences with values g for distance d and angle θ, and $H_{\theta,d}$ is the total number of grey level differences.

Entropy describes the degree to which distinct scales of organization are unrecognizable. It is maximum when all radiance differences have an equal probability of occurring (i.e. the histogram is uniform) and low when texture is smooth.

$$ENT(\theta,d) = \sum_g \frac{h_{\theta,d(g)}}{H_{\theta,d}} \log \frac{h_{\theta,d(g)}}{H_{\theta,d}} \qquad 2$$

Hobson (1972) utilizes a measure called vector strength. If the pixels within a cell are connected into a set of adjacent triangular planes, then texture can be measured through the dispersion in three-dimensional space of normal vectors to these planes. Vector strength is a summary of the distribution of normal vectors and is high for smooth surfaces and low for rough surfaces. See Hobson (1972) or Key (1990) for details.

These texture measures are calculated for ten cloud classes which include some of the basic cloud groups and mixtures of these as observed in the data:

1: Low thin cloud over water (stratus);
2: Low thin cloud over ice (stratus);
3: Low thin cloud over land (stratus);
4: Low thick cloud, smooth texture (stratus);
5: Low thick cloud, bumps or broken (stratocumulus);
6: Middle cloud rolls (broken, linear altostratus usually over a stratiform layer);
7: Broken middle cloud, not linear;
8: Middle thick cloud, smooth (altostratus, possibly over stratus);

9: Middle/high bumps (cirrocumulus or altocumulus);
10: High thick cloud with some middle cloud (broken cirrostratus over altostratus);

The surface was included in classes 1-3 only because the clouds are thin and differed primarily in albedo. Contributions from surfaces to cloud albedo or temperature in the other classes were not significant enough to justify defining additional classes. Class 7 is similar to class 6, but occurred at a higher altitude (lower temperature).

8.2.2 Results

Unlike the previously-described cloud and surface classification, this example uses local textural as well as spectral characteristics. The training set comprised approximately five percent of the image area; the same data were used for both the NN and ML approaches. The classification results indicate that, as expected, cloud fields are organized into recognizable mesoscale morphologies although their quantification remains a problem. The overall classification agreement between the ML results and the manual classification is 68% with 10% of the image left unclassified. The largest differences are due to differences in the location of boundaries between cloud systems, labeling of low thin cloud over ice as low thick cloud by ML (Kara Sea), ML detecting a linear pattern in some middle cloud areas which appeared broken in the manual interpretation, and some low thick cloud areas labeled as middle broken by ML.

The neural network classification compared favorably with the ML results where the classification agreement between them was approximately 80%. Agreement with the manual interpretation was only 60%, with inaccuracies similar to those for the ML/manual comparison. It can be seen that the classification results could be improved by redefinition of these classes and by choosing more appropriate training areas. However, given the complexity of the problem - how to define and classify cloud patterns - more information is gained by studying why this method performed poorly than by tuning the classifier simply to achieve higher classification accuracy.

8.3 Discussion

The patterns of weights in the neural networks provide insight into the way decisions are made by the network. For the example using spectral characteristics to classify surface and cloud types (Section 8.1) in particular, interpretation of these weights sheds light on which spectral channels are important in the identification of these classes. Similarly, the network weights indicate, for each output class, the degree of information redundancy among channels in the input data; channels that are only weakly weighted add little additional information to the classification process. Figure 8.3 shows an example of the connections between the input channels, hidden layer neurons, and the output classes in the trained network. Due to the complexity of the connections between units, only the surface classes are shown in the figure. The identification of the exact role of hidden units is difficult, as they often represent generalizations of the input patterns. The strength of the connections shown varies from 0.6 to 0.9 (on a scale of -1.0 to +1.0). These connections are summarized in Tables 8.2 and 8.3. Table 8.2 shows which input data channel each hidden node is associated with in the trained network.

Table 8.2: Connections between the input channels and the hidden layer in the trained neural network.

Hidden Layer (Neuron Number)	Input Chanel Connections (AVHRR: 1,2,3,4,5; SMMR: 18,37)
0	1,2,4,5
1	NONE
2	5,18,37
3	4,5,18
4	2,18
5	NONE
6	4,5,18,37
7	2,3
8	1,18
9	2,3,4,5

Table 8.3: Connections between the output class and the hidden layer in the trained neural network.

Output Layer (Class)	Hidden Layer (Neuron Number)
LAND	2,3,6
SNOW	0
WATER	1,3,5,6,7,8,9
ICE	1,3,5,8
LCLL	6
LCLW	0.9
LCLI	9
MCLW	1,4,5,7,9
MCLI	0,4,5,9
HCLL	2,4,7
HCLW	1,4,5,7,8
HCLI	0,3,4,8

Table 8.3 shows the association between hidden units and the output data classes. Following the connections through these two tables therefore indicates which

input channels are linked to particular output classes. As shown in Figure 8.3 and Table 8.3, snow-free land has strong connections with hidden layer neurons 2, 3, and 6, all of which represent thermal AVHRR channels and the SMMR channels (Table 8.2). We may therefore conclude that land is best distinguished from the other channels by its physical temperature and its emissivity in the microwave portion of the spectrum. Snow/ice cap is identified by its albedo and temperature, with no significant information gained from the microwave signature.

The identification and discrimination of ice from the other classes requires albedo, temperature, and microwave emissivity characteristics. The connections show that cloud identification is a function of height, with thermal characteristics being more important for middle- and high-level clouds. The identification of low cloud depends on the underlying surface, where temperature is an adequate discriminator if the cloud is over land, albedo and temperature are used if over water, and temperature and the longer-wave reflected solar component (AVHRR channel 3) are needed if over ice. The hidden layer neurons with connections to AVHRR channel 3 (numbers 7 and 9) are also connected to either channel 2 or channels 4 and 5, indicating that both the reflected solar and thermal components of channel 3 likely play a part in the classification.

The connections demonstrate the usefulness of AVHRR channel 3 for discriminating between cloud and snow or ice. Finally, note that hidden layer neurons 1 and 5 do not 'listen' to any inputs, and therefore do not add any information to the network. It should be pointed out, however, that while the connections just discussed have reasonable physical interpretations, this is not always--or even often--the case. Typically there seems to be no unambiguous explanation for their weights.

Is it useful to compare classification results from the neural network and maximum likelhood methods? Although both classifiers use the same training data, some fundamental differences exist in the way they are used. The neural network does not directly address the mean and covariance within a training area. Instead, each pixel within the training area is a separate pattern that directly influences the development of node weights. The multispectral characteristics of each pixel imprints itself to some degree on the network connections. During the development of unit weights as part of the network training phase, some aspects of training area means and covariances are included in the weight assignments. However, unlike the ML classifier, the neural network is not limited to assuming a statistical relationship between pixels within a class and is not restricted by assumptions of normality in the data. The fact that the multispectral data used here often violate these assumptions may contribute to the low percentage of the data classified using ML with the training statistics in TA1.

To test this hypothesis, a synthetic data set of AVHRR and SMMR data was developed that exhibited truly normal data distributions. No attempt was made to generate realistic cloud texture. Rectangular cloud and surface objects of varying sizes and locations were created whose dimensions were randomly chosen within a restricted range. Object regions were then filled with normally distributed data for each channel based on pre-specified means and standard deviations (Gaussian random number generator) characteristic of the polar clouds and surfaces. With this data set there will be only one statistical class for each physical class (e.g., land, low cloud over water, etc.). For this reason, and because the data are normally distributed with a known variance, the probability of selecting a training area representative of the population is higher than with the actual data. Therefore, even small training areas should provide enough information about each class to allow a larger proportion of the area to be classified. This was in fact the case, where training areas extracted from less than 1% of the synthetic image allowed approximately 70% of the image to be correctly classified by the ML procedure. This test suggests that deviations from a normal distribution likely contribute to the low percentages of classification using the ML classifier.

The ability of a neural network to compute similarity measures through a comparison of patterns contributes to its ability to classify large portions of data on two separate images. Thus, although a relatively small portion of the variability of clouds and surfaces were captured in the training areas (particularly TA1), the neural network was still able to reliably choose the most appropriate output class. This property provides a means to address the problem of signature extension over time and space, since a properly trained network can make class assignments - albeit with reduced confidence - in the face of atmospheric effects or slight changes in spectral properties without requiring apriori knowledge of within-class variance or probabilities of class membership. In fact, if one has a particular reason to use a statistical classifier, the strength-of-membership values calculated by a neural network could be fed back into the statistical classifier as *a-priori* probabilities.

The classification examples presented have utilized numeric data--albedos, temperatures, textures--as input. Output is on the nominal level for both the neural network and the ML procedure, although the neural network also provides a type of membership value. In some cases, input such as category identifiers rather than measurements may be useful where pixels are assigned a class symbol and optionally an associated fuzziness value (e.g., the probability that the pixel belongs to the class). Another neural network was developed that uses both nominal and categorical input, again with the purpose of classifying cloud and surface types without texture features; i.e., as in Section 8.2. For example, in the study area the locations of land and permanent ice cap are known, and the location and concentration of sea ice can be determined from the SMMR data. Consider the case where only three broad categories of sea ice concentration are of interest: low (15-40%), medium (41-70%), and high (71-100%). Other variables are also possible; for example, time of year, geographic location, spatial context, texture, stage of plant growth, and the a priori probability of occurrence of each surface or cloud type. For simplicity, however, the example is limited to symbols representing land/not land, ice cap/not ice cap, and low, medium, and high sea ice concentration variables. This network was trained with input units corresponding to both spectral and categorical variables. Since some of the AVHRR and SMMR channels are highly correlated - as evidenced from principal components analysis and an examination of the previously described neural network, only AVHRR channels 1, 3, and 4, and the SMMR, 18 GHz vertical channel were used in the training. Categorical input variables represent land, ice cap, and sea ice concentration. Nine hidden layer units were specified with output classes as before. As expected, the resulting classification (not shown) is similar to that using only spectral information but the proportion of correctly identified surface pixels increased slightly, whereas the proportion of cloudy pixels remained essentially the same. In addition, the certainty with which surface pixels were classified as measured by the output membership values increased significantly; with some coastal pixels the increase was as much as 0.4.

8.4 Other Neural Network Applications to Cloud Classification

The two examples provided thus far are not the only applications of neural networks to cloud classification. Others that are relevant to the research presented in this chapter are described below. The overall goals and approaches of these studies are similar to that presented in Section 8.1; i.e., classifying pixels into one of a number of cloud and surface types using a neural network. They differ, however, in the number and types of classes, the input features, and the type of neural network employed.

Rabindra *et al.* (1987) used a hybrid neural network/expert system to classify polar scenes using AVHRR Local Area Coverage (LAC; 1.1 km pixel size at nadir) data. Ten classes were of interest: open water, solid sea ice or snow-covered land, broken sea ice, snow-covered mountains, thin stratus or fog over ice, cirrus over ice, stratus over water, cumulus over water, multi-layer clouds, and snow-free land. A total of 183 spectral and textural signatures were examined although only 20 were selected for use in the classifier. The expert system component was designed primarily as a convenient user interface for training and testing. The 'inference engine' was a probablistic neural network (PNN). The PNN is similar in structure to a back propagation network, but the sigmoid function is replaced by one of a class of functions which include, in particular, the exponential. One important advantage to the PNN is that training requires only a single pass. After every sample cell in the image is classified by the PNN, an outlier test is applied to check if that sample really belongs to the selected class. The approach was very successful with an overall classification accuracy of 87% for the images tested.

Welch *et al.* (1987) compared three different classifiers for the same task as Rabindra *et al.* (1987), that of classifying arctic scenes in AVHRR LAC data. The same 10 classes and 20 features were used. Two of the three classifiers were neural networks: a forward-feed back-propagation architecture (FFBP) and the probabilistic neural network. The other classifier was the traditional stepwise discriminant analysis. The overall accuracies obtained by the methods for six AVHRR images were 87.6%, 87.0%, and 85.6% for the FFBP neural network, the PNN, and discriminant analysis, respectively. Thin cloud was the class with the lowest classification accuracy (approximately 75%) for all the classifiers. This work was extended by Tovinkere *et al.* (1992) to include three addition neural network variants as well as a fuzzy logic-based expert system.

8.5 Sea Ice Fracture Patterns

We have so far focussed on the retrieval of cloud cover from satellite data. As discussed in the Introduction, however, sea ice plays a crucial role in the complex interaction of ocean and atmosphere. For example, a reduction in the extent and thickness of sea ice due to global warming, and the consequent increase in the number of cracks in the ice (hereafter 'leads'), is expected to further increase global temperatures. This positive feedback is a result of reduced albedo and the increase in heat transfer from the ocean to the atmosphere. Model estimates indicate that an increase of 4% in the area covered by leads during winter could produce a hemisphere-wide warming of 1 degree Kelvin (Ledley, 1988). Understanding lead formation processes as well as the geographical and temporal distribution of lead networks is, therefore, important to studies of global climate. Unfortunately, no reliable automated methods for doing so have been developed, in part due to the difficulty of defining leads quantitatively. In this section a neural network approach to the identification of leads in Landsat data is examined. The specific tasks to be performed by the networks are to:

1: remove any trends and significant noise in the images,
2: identify pixels that belong to leads,
3: group those pixels into lead fragments, and
4: group the lead fragments into lead networks.

To accomplish these goals, an expert system was initially developed (Key *et al.*, 1986). A neural network was then developed to achieve the first three goals above, and another network was developed to address the last item. While our focus is on the neural network approaches, a brief description of the expert system will aid in illustrating the problem. Once again, it is important to keep in mind that while the applications are to sea ice features, the general procedures apply equally well to other linear or curvlinear features such as stream channels or roads.

8.5.1 Lead Retrieval Using a Knowledge-Based System

Expert (or knowledge-based) systems consist of a set of facts about the state of the system and a group or groups of if-then type rules that perform actions to change the system state. They differ from the traditional algorithmic style of programming (e.g., using FORTRAN) in that the rules are not necessarily evoked (or 'fired') in a predictable way since they are driven by the data in the knowledge-base and in turn may change it. For that reason they are extremely flexible and provide for rapid prototyping. On the other hand, simulating the flow of data through an expert system can be somewhat difficult.

Knowledge-based systems have been developed for the classification of aerial photos and satellite images (e.g., Nazif and Levine, 1984; McKeown *et al.*, 1985; Matsuyama, 1987; Matsuyama, 1989; Schowengerdt and Wang, 1989; Skidmore, 1989) although their use in the detection and mapping of linear features has been restricted to well-defined features such as roads, rather than to objects whose spectral and geometrical properties vary widely like the leads considered here (e.g., Stansfield, 1986; Niblack *et al.*, 1988; Wang and Newkirk, 1988). As will be shown in this and subsequent sections, the problem is a difficult one and the approaches have met with only limited success.

8.5.2 Methodology

Algorithm development and testing is based on a Landsat-5 MSS band 5 (approximately 0.6 μm) image of the ice pack north of Alaska in the Beaufort Sea (76°N, 137°W). The image was acquired on March 12, 1988 and covers an area of approximately 10,404 km^2. A transparency of this MSS scene was scan-digitized to yield digital data with 256 grey levels and a pixel size of 200m. The study area is shown in Figure 8.5. Present in the image are open leads, refrozen leads, and sea ice. The complexity of the problem is obvious: some leads are dark (thin ice) others are relatively light grey shades (thicker ice), and there are many discontinuities in the lead network. The goal is to identify the leads in the image and map the network, seeing through the noise and connecting the fragments. The overall procedure consists of three steps:

1: The original image is mapped to a lead/not-lead binary image with a dynamic threshold method and initial lead fragments are located with a region-growing procedure. Processing at this level is entirely algorithmic.
2: A set of rules are applied to generate hypotheses concerning which fragments are likely to be real leads and which are connected. Numeric procedures, invoked as rules, are fired to determine various statistical and geometric properties of the fragments (or 'objects'). Initial statistics are extracted for valid objects.
3: Pairs of valid objects are examined in terms of their geometrical relationships, and are connected where possible. Processing stops when no more fragments can be merged and all statistical characteristics have been determined.

Image processing routines were integrated with the C Language Integrated

Production System (CLIPS). CLIPS is a rule-based system development tool produced in 1986 by the Artificial Intelligence Section at the NASA Johnson Space Flight Center. It was designed to provide high portability, low cost, and easy integration with external systems. The primary representation methodology is a forward chaining rule language based on the Rete algorithm. CLIPS rules can call external procedures which can in turn modify the knowledge base. Thus, data can be passed both to and from the CLIPS environment.

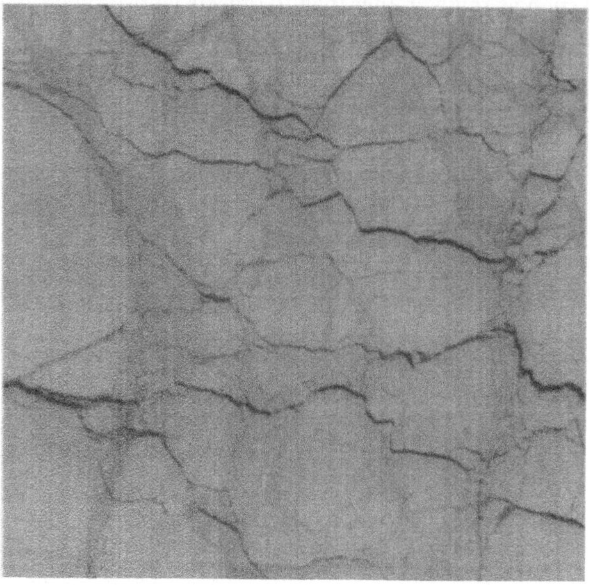

Figure 8.5: Landsat MSS band 5 (visible) scene of the Beaufort Sea on March 12, 1988.

The first processing step is to classify the original image into those pixels that are part of a lead (or presumed to be so) and those that are not lead pixels. Assumptions in this procedure are that leads are less reflective than the surrounding ice, and that clouds are not present. Given these conditions, only two classes should be present in the original image. Unfortunately, sensor noise, varying surface conditions, and illumination trends in the image due to the low sun angle are complicating factors which preclude the use of a single threshold value. Therefore, a modified version of the minimum error method of Chow and Kaneko (1972) was employed to determine thresholds dynamically throughout the image.

In this procedure, the image is divided into small regions and a histogram is computed for each. These are examined for bimodality, under the assumption that such a histogram is an estimate of the probability density function of a mixture population comprising the grey levels of the object and background pixels. It is assumed that each of the two components is normally distributed with a unique mean, variance, and *a-priori* probability. The problem is then one of locating such histograms within the image, estimating these parameters, and then determining - through the method of maximum likelihood - the threshold value of each bimodal distribution that minimizes the probability of misclassification. The population parameters can be estimated by the criterion function given in Kittler and Illingworth (1986). This and other thresholding methods are summarized in Sahoo *et al.* (1988).

In this example, histograms for 8x8 pixel cells were examined for bimodality. In the Chow and Kaneko (1972) study, cells overlapped thereby increasing the total number for which thresholds are calculated. Only 30 of the cells were found to have bimodal distributions satisfying the above criteria; from these an image of thresholds was created, filling empty cells with the nearest threshold. This method, as shown later, proved sufficient although a more sophisticated interpolation scheme may prove to be more robust. From the binary image of lead/not lead pixels, a simple region-growing procedure identified and labeled each group of connected lead pixels. These initial lead objects were then examined in terms of their geometric properties and statistical characteristics.

Processing the binary image consists of the following steps:

1: small objects are merged with neighboring objects or discarded if isolated;
2: remaining objects are described in terms of linearity, where the linear correlation coefficient, r, is used as the measure of linearity. If r is statistically significant, the slope of the regression line is used as a measure of orientation. Those objects whose r is not significant are left undetermined;
3: lead widths, lengths, and orientations are computed for each valid object;
4: pairs of valid objects are examined in terms of proximity and orientation and may be merged. Orientations are recomputed for each new object.

Four rules were used to discard objects that are too small and/or too far from valid objects. For each object, a fact is asserted containing information on its size, location, and linearity. One rule examines this information and, if the object is very small, tries to merge it with a neighbor. If merged, the fact for the smaller object is retracted and a new fact is asserted for the merged object. This causes other rules to recompute the geometric and statistical properties of the object. When no more objects can be merged at this level, the procedure to retrieve the width, orientation, and length information for each valid object is called. The next step is to connect valid objects, and a new goal is asserted.

Valid objects are connected in order to obtain information on the larger-scale objects to which they belong. With leads and lead networks, however, this becomes a problem of definition. Leads are extremely dynamic; fractures may occur in any direction, and leads begin to refreeze immediately after forming. In short, there is no accepted objective procedure for mapping leads. Here we connect objects that are in close proximity and have similar orientations, using CLIPS rules to check these features and act accordingly. Other rules will then fire to make the connection and recompute statistics.

8.5.3 Results

The binary image resulting from the application of the dynamic thresholds is shown in Figure 8.6. Also shown in Figure 8.6 are the valid lead objects derived from the binary image, where intermediate grey shades represent lead fragments that were rejected due to their size or shape. Figure 8.7 gives relative frequencies of the lead widths and orientations (as slope) for these objects. The results are physically realistic; for example, the lead width distribution shows a tendency towards a negative exponential distribution, which is the mathematical distribution expected if lead widths are random. Additionally, the distribution of orientations shows a realistic preferred orientation of major leads. Valid objects that were connected by the rules described above are shown in Figure 8.8. While the results are not terribly impressive, this preliminary mapping scheme is capable of summarizing useful physical patterns in the lead complexes.

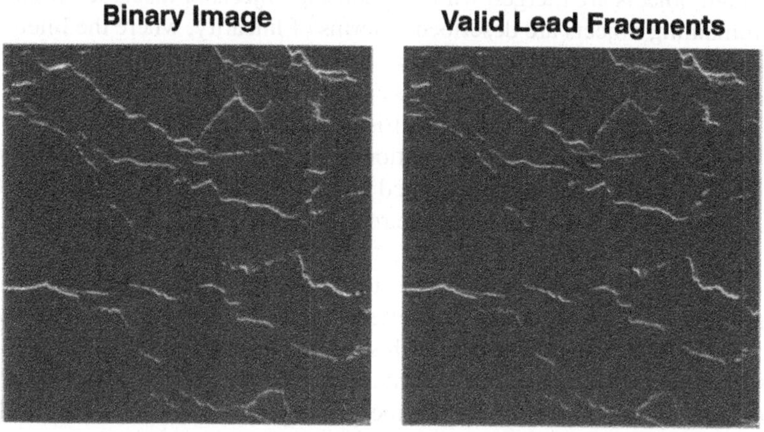

Figure 8.6: A binary version of the image in Figure 5 (left) created using the dynamic threshold method, and the valid lead fragments (right) derived using the rule-based system from the binary image.

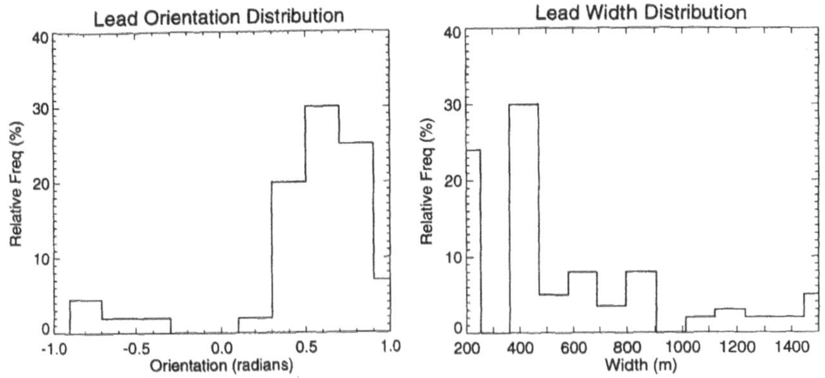

Figure 8.7: Lead orientation and width distributions for the lead fragments that are shown in Figure 8.6.

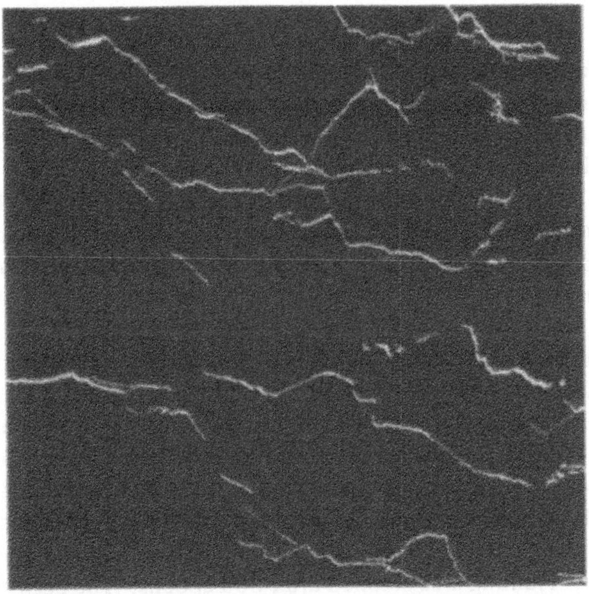

Figure 8.8: Connections between the lead fragments of Figure 8.5 as determined by the rule-based system using orientation, size, and proximity information.

8.6 Neural Network Approach

In this section two neural networks are described that deal with the lead retrieval problem. The first network was developed for the purpose of creating a binary image from the original grey-scale image, thereby replacing the dynamic threshold procedure described in the previous section. The second network is used to identify and connect associated lead fragments in the binary image through a cooperative-competitive procedure.

8.6.1 Grey-Scale to Binary Image Transformation

The neural network learning and processing technique for mapping a grey scale image into a binary one is the feed-forward network using standard back-propagation learning. Each training input - a small window in the original satellite image - is represented as a pattern of activity on the input units. The target output for the output units is the binarized version of the input, determined through a manual interpretation. The network consisted of 120 hidden units, 64 input units and 64 output units (8x8 windows).

Examples for training the network were obtained from the image shown in Figure 8.9. The binary image was obtained by manually thresholding the image. This manual thresholding yielded a slightly better image than that that obtained with dynamic thresholding. Small windows of 8x8=64 pixels were cut from the images. These, scaled to a range of [0.2, 0.8], constitute the input and target patterns. Scaling between 0.2 and 0.8 (instead of 0 and 1) constrained the activations to lie within the linear portion of the sigmoid, which was found to ease learning. Various experiments were run with training patterns of 64 pixels (widows of 8x8 pixels) and 256 pixels (widows of 16x16 pixels).

Since the binary image from which the target patterns were obtained is far from perfect (Figure 8.10), the training set on which the network learned was only a small subset of all possible training patterns. Target training windows were manually selected. Input/target pairs of patterns selected were furthermore constrained to have a limited number of lead pixels, 50 in most experiments. The network was trained on 300 patterns each containing at least one lead pixel. An additional 481 patterns corresponding to areas with no leads were added to that training set. The learning rate was 0.05 and the momentum was 0.9. Training was stopped after 600 epochs. The network simulator available in McClelland and Rumelhart (1988) was initially used. Figure 8.11 shows the total squared error as a function of the number of training epochs. The number of incorrectly classified pixels is plotted with the dashed line. Figure 8.12 shows the output of the network for the training image, where each output unit was thresholded at 0.5. A number of other experiments with different numbers of hidden units and other selections of input patterns were also performed and yielded output images with various degrees of noise, but that were qualitatively similar to that shown in Figure 8.12.

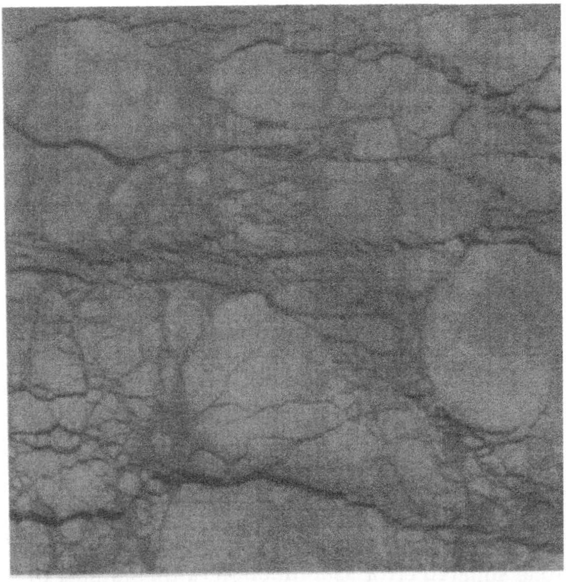

Figure 8.9: Another Landsat image of the Beaufort Sea showing a network of leads.

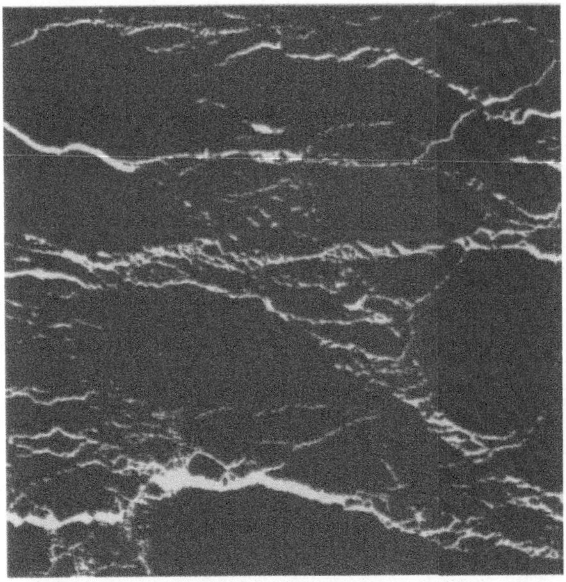

Figure 8.10: Binary image of Figure 9 created by manual thresholding.

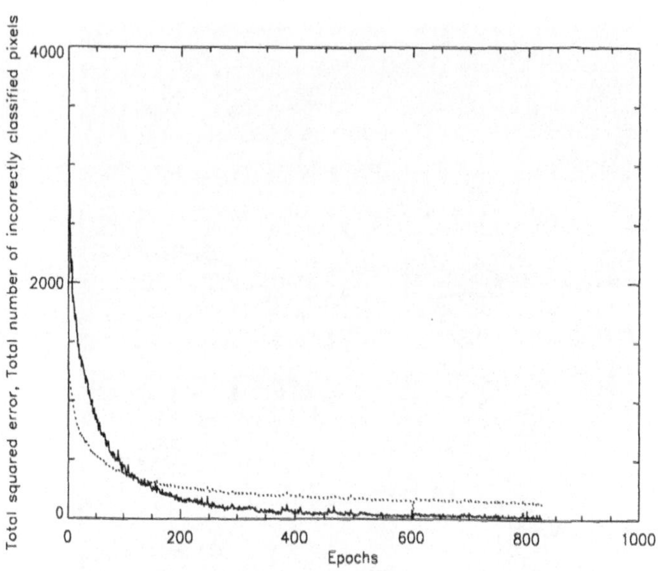

Figure 8.11: The total squared error as a function of the number of training epochs. The number of incorrectly classified pixels is also shown (dashed).

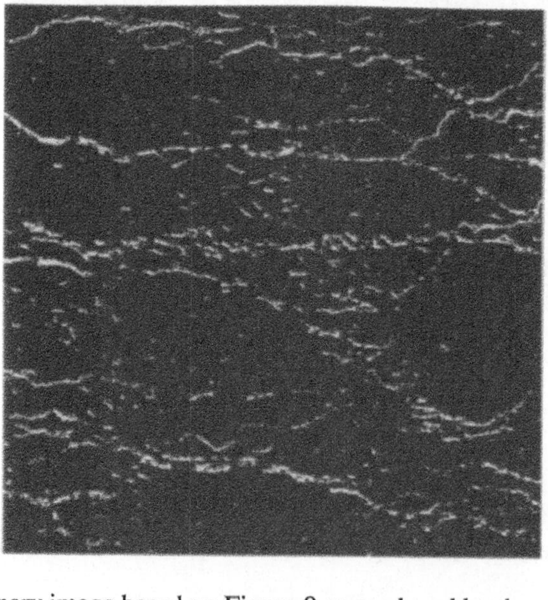

Figure 8.12: Binary image based on Figure 9 as produced by the neural network.

While the results obtained are far from optimal, and more experiments need to be run on other images, the results are encouraging: the networks, which were 'generic', were able to binarize a new unseen image. Far better results can hopefully be obtained with an architecture specifically designed for the task at hand. These preliminary experiments also show that the networks' behaviors were sensitive to a number of factors:

1: Obtention of the binary image used for training. The networks are trained on a less than optimal target image, as it was thresholded manually on only one cut-off point. One alternative considered to this somewhat ad hoc strategy is to use a bitmap editor and manually edit the original image, drawing the leads. This is rather time consuming, but worthwhile, since this needs to be done only once.

2: Selection of training examples containing lead data. As expected, the large majority of 8x8 windows within the binary image contain no lead pixels. Since the networks used could not learn all available training patterns, a limit was imposed on the number of training patterns chosen for each given number of lead pixels in the target image. Performance was found to be highly sensitive to this selection. A better preliminary study of pattern distribution, possibly in terms of entropy, might yield an improved understanding of how the best selection of examples can be achieved. Surely, the use of a neural network constrained to process patterns with shift invariance with respect to lead positions would yield better results.

3: Selection of training patterns corresponding to no leads. Two types of noise, random and organized, are present in the satellite image. Random noise is due to cloud cover and imprecision in the recording process, and consists of small specks on areas usually limited to a few pixels width. Organized noise, like the brighter vertical or horizontal lines clearly visible in the images, are probably originating from scanning devices used in the photographic process. To prevent the network from treating these lines as leads, and to prevent the network from treating any dark noise as a lead segment, it is crucial to present it with counter examples where no leads are detected.

4: Regularization through overtraining. All experiments showed that performance was better when the network was overtrained, that is, when the number of training patterns was high and learning was stopped before the error was too low.

These experiments used a network that was not constrained to process patterns in a shift invariant manner. Doing so, however, would undoubtly improve generalization, and potentially speed up learning. Such a constraint can easily be added in a back-propagation network, at a cost of a slightly more complex architecture, by using a layered and modular architecture where certain weights are shared. This type of network has not yet been tested.

8.6.2 Joining Lead Segments

As an alternative to the rule-based system previously used to join lead fragments, a cooperative-competitive neural network has also been explored. The goal is to connect lead fragments that probably were at one time part of the same lead, but are disconnected in the image. 'Probably' means that they have similar orientations and are in close proximity. This type of problem has received some attention in the literature, although not for the mapping of sea ice leads. Applications to other geophysical variables are certainly possible; for example, roads or stream systems.

Figure 8.13 shows the basic architecture of the network. In this design dark pixels representing leads reinforce each other, competing against mutually reinforcing white pixels representing ice. The orientation of lead pixels is computed and used as input to the network. For each cell in the image there is a pool of units, with each unit representing a different orientation (bottom level in Figure 8.13). Units within a pool are connected, and pools representing 8 image cells (up, down, right, left, and on the diagonals) are also connected. Each input cell then activates each unit in its pool, with the activation level prescribed according to a bell (Gaussian) function. The more similar the orientation of a cell is to a particular unit, the higher its activation level will be. Short-range competition takes place within each pool resulting in each unit having an

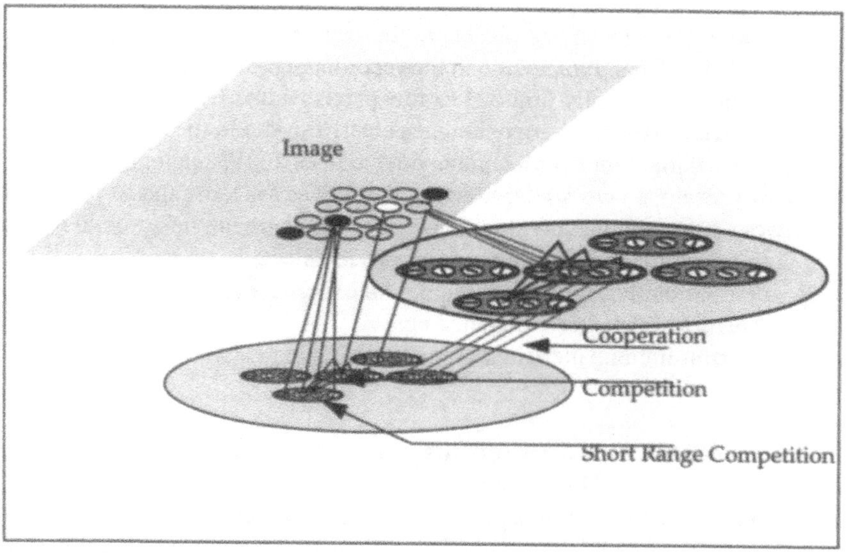

Figure 8.13: The basic architecture of a competitive-cooperative network for joining lead fragments.

activation level of either 0 or 1. Competition between the pools takes place (accomplished using negative weights) so that the dominant orientation within a group of cells emerges. At this stage the orientation of the lead within each pool is modified, being 'pushed' toward the average orientation of the group of cells.

Cooperation between the pools (accomplished using positive weights) then works to relabel those cells that have no obvious orientation (i.e., no lead) based on the modified orientations of the cells around it. The original pixel values for all cells that exhibited definite orientations are copied to the output image, while the pixels within cells that had no obvious orientations are relabelled based on the average orientation determined during competition. In other words, the lead cells cooperate in filling in the empty cells between them.

The algorithm was first tested on some simple artificial images, and proved to be successful in those cases. Figure 8.14 shows two testcases, with images on the left as input, and images on the right showing the results. The network was then used with a real image. Again, a Landsat image was used in the analysis. Part of one lead taken from the binary version of the image is shown in Figure 8.15. The resulting 'complete' lead is also shown. While the results for this one lead are promising, in general the method is very sensitive to cell size, orientation and training time. Additionally, the complexity and computational costs of the implementation of such a network, as well as the large amount of noise present in the original images, are factors to be considered when evaluating the utility of such an approach.

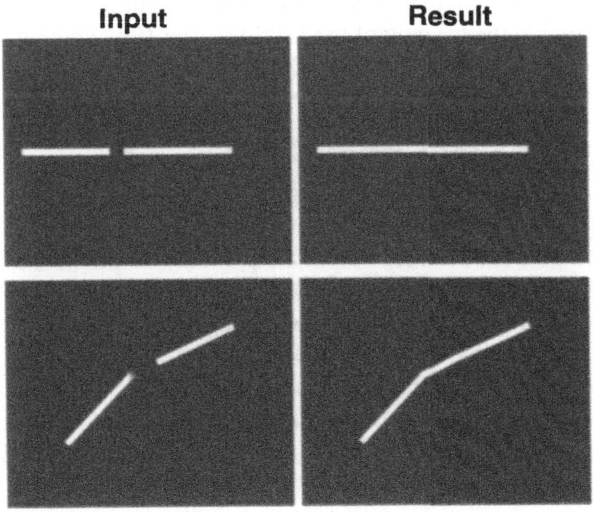

Figure 8.14: Test images (left) for the competive-cooperative network used to join lead fragments (right).

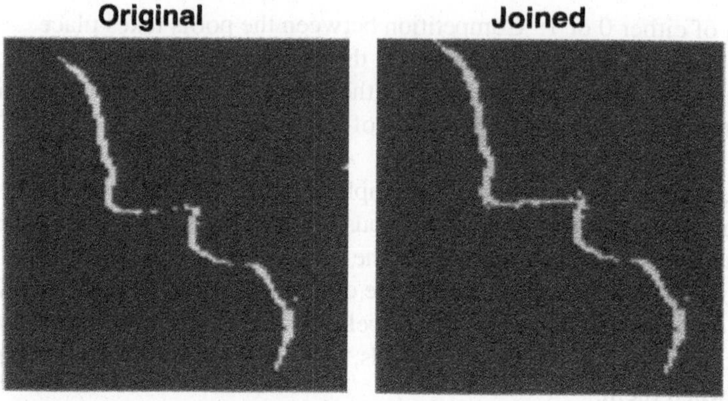

Figure 8.15: Part of a lead in a Landsat image (left) used to test the competitive-cooperative network for joining lead fragments. The results are shown on the right.

8.7 Conclusions

In the classification of broad surface and cloud types in satellite data, the neural network approach is generally less rigid than the traditional maximum likelihood procedure in that 1) there are no assumptions about the statistical distributions of variables and relationships between them, 2) the network is easily trained to learn the relationships between input and output, and 3) the classification produces both a categorical value and a type of membership value for each pixel. It is recognized that there is some loss of information and interpretabilty with the departure from statistical theory. Additionally, the computation time required for training the network is not trivial when compared to the training of the ML classifier (i.e., computation of mean vectors and the covariance matrix), although future hardware architectures should alleviate this problem.

The ability to interpret weights within the trained network provides a potentially powerful tool for understanding the role of inputs and the geophysical processes they represent in the making of decisions. Through an examination of the connection strengths between input, hidden, and output units, it may be possible to identify which inputs influence the classification most, and which are redundant. These relationships are not always clear, and care must be taken in extending their interpretation to physical processes. It was also shown that ancillary information, even on a simplistic level, can improve classification accuracy and can be easily included in a network. Although the cloud classification example indicated that maximum likelihood results could be made to agree more closely with the manual interpretation, this was achieved only after training areas were expanded to include 9% of the test image. Such a degree of training is impractical for remote sensing climate studies because of the volume of imagery that must be processed.

The knowledge-based systems approach to lead detection, parameter extraction, and mapping taken here incorporates both numeric and symbolic processing schemes. This integrated system has several design advantages. In particular, the rule-based environment is generally a more flexible approach to specification of the problem, which in this case is the detection and mapping of features that are linear on the small (10 km) and large scale (100 km), but often curvilinear and disjointed when examined on the medium scale (30-60 km). Because this type of problem is really one of defining what a lead actually is, the definition-testing cycle occurs many times and includes a high degree of subjectivity. The rule-based approach thus allows the direct implementation of domain knowledge as it becomes available. Additionally, this framework allows for rapid prototyping, in part because processing steps can be specified in broad terms, and cooperating processes are more easily implemented than in the usual sequential mode of execution. The neural network counterparts to the knowledge-based system for image binarization and analysis of linear features also show promise. The examples given were, unfortunately, particularly sensitive to image noise and cell size so that further research is warranted.

Image processing using neural networks will undoubtedly become more common in the coming years. The disadvantages of this approach are that training time is significant, the network operates essentially as a 'black box' (but see Chapters Six and Seven), and the procedure is not based on physical principles; e.g., the physics of radiative transfer are not directly represented. As we have seen, the advantages are that the networks are relatively easy to construct, very flexible, and do not require a complete physical understanding of the system. In cases such as those presented in this chapter, the advantages of neural networks for feature extraction outweigh the disadvantages.

Acknowledgements

This work was supported by NASA grant NAGW-1786. Thanks are due to J. Maslanik and A. Schweiger for their efforts in the analysis of clouds presented in Section 8.1, and to O. Brousse who preformed the analysis in Section 8.6.

References

Barry, R. G., Henderson-Sellers, A. and Shine, K. P. (1984) "Climate Sensitivity and the Marginal Cryosphere", in Climate Processes and Climate Sensitivity, edited by J. Hansen and T. Takahashi, Geophys. Monogr., Vol. 29, AGU, Washington, D.C., pp. 221-337.

Borchardt, G.C. (1986) "STAR: A Computer Language for Hybrid AI Applications, in Coupling Symbolic and Numerical Computing in Expert Systems", J.S. Kowalik (ed.), Amsterdam, The Netherlands: North-Holland.

Campbell, W.J. and Roelofs, L.H. (1984) "Artificial Intelligence Applications for the Remote Sensing and Earth Science Community", Proceedings, Ninth Pecora Symposium on Spatial Information Technologies for Remote Sensing Today and Tomorrow, Sioux Falls, South Dakota.

Chow, C.K. and Kaneko, T. (1972) "Automatic Boundary Detection of the Left Ventricle from Cineangiograms", Computers and Biomedical Research, 5, 388-410.

Committee on Global Change (1988) "Toward an Understanding of Global Change", National Research Council, Washington, D.C.: National Academy Press.

Estes, J.E., Sailer, C. and Tinney, L.R. (1972) "Applications of Artificial Intelligence Techniques to Remote Sensing", Professional Geographer, 38 (2), 133-141.

Hobson, R.D. (1972) "Surface Roughness in Topography: Quantitative Approach" in Chorley, R.J. (ed.), Spatial Analysis in Geomorphology, New York: Harper & Row, 393 pp.

Jackson, M.J. and Mason, D.C. (1986) "The Development of Integrated Geo-Information Systems", Int. J. Rem. Sens., 7 (6), 723-740.

Key, J. (1990) "Cloud Analysis with Arctic AVHRR Data, Part 2: Classification with Spectral and Textural Features", J. Geophys. Res., 95(D6), 7661-7675.

Key, J. and Barry, R.G. (1989a) "Cloud Analysis with Arctic AVHRR Data, Part 1: Adaptation of the ISCCP Cloud Detection Algorithm", J. Geophys. Res., 94(D15), 18521-18535.

Key, J., Maslanik, J.A. and Schweiger, A.J. (1989) "Classification of Merged AVHRR and SMMR Arctic Data with Neural Networks", Photogram. Eng. Remote Sens., 55 (9), 1331-1338.

Key, J.R., Maslanik, J.A. and Barry, R.G. (1989b) "Cloud Classification Using a Fuzzy Sets Algorithm: A Polar Example", Int. J. Rem. Sens., 10(12), 1823-1842.

Key, J., Schweiger, A.J. and Maslanik, J.A. (1986) "Mapping Sea Ice Leads with a Coupled Numeric/Symbolic System", ACSM-ASPRS Proceedings, Vol., 4, 228-237.

Kittler, J. and Illingworth, J. (1986) "Minimum Error Thresholding, Pattern Recognition", 19, 41-47.

Kitzmiller, C.T. and Kowalik, J.S. (1987) "Coupling Symbolic and Numeric Computing in KBsystems", AI Magazine, 8 (2), 85-90.

Kowalik, J.S. (ed.) (1986) "Coupling Symbolic and Numerical Computing in Expert Systems", Amsterdam, The Netherlands: North-Holland.

Lauritsen, L., Nelson, G.G. and Port, R.W. (1979) "Data Extraction and Calibration of TIROS-N/NOAA A-G Radiometer", NOAA Tech. Memor., NESS 107, Natl. Oceanic and Atmos. Admin., Boulder.

Ledley, T. (1988) "A Coupled Energy Balance Climate-Sea Ice Model: Impact of Sea Ice and Leads on Climate", J. Geophys. Res., 93 (D12), 15919-15932.

Maslanik, J.A., Key, J.R. and Barry, R.G. (1989) "Merging AVHRR and SMMR Data for Remote Sensing of Ice and Cloud in Polar Regions", Int. J. Rem. Sens., 10(10), 1691-1696.

Matsuyama, T. (1989) "Expert Systems for Image Processing: Knowledge-Based Composition of Image Analysis Processes", Comp. Vision, Graphic and Image Proc., 48 (1), 22-49.

Matsuyama, T. (1987) "Knowledge-Based Aerial Image Understanding Systems and Expert Systems for Image Processing", IEEE Trans. Geosci. Rem. Sens., GE-25 (3), 305-316.

McClelland, J. L. and Rumelhart, D. E. (1988) "Explorations in Parallel Distributed Processing: Computational Models of Cognition and Perception." The MIT Press.

McKeown, D.M., Harvey, W.A. and McDermott, J. (1985) "Rule-Based Interpretation of Aerial Imagery", IEEE Trans. Pattern Anal. Mach. Intell., PAMI-7 (5), 570-585.

Nandhakumar, N. and Aggarwal, J.K. (1985) "The Artificial Intelligence Approach to Pattern Recognition - A Perspective and an Overview", Pattern Recognition, 18 (6), 383-389.

Nazif, A.M. and Levine, M.D. (1984) "Low Level Image Segmentation: An Expert System", IEEE Trans. Pattern Anal. Mach. Intell., PAMI-6 (5), 555-577.

Niblack, W., Petkovic, D. and Damian, D. (1988) "Experiments and Evaluations of Rule Based Methods in Image Analysis", Proceedings: IEEE Conference on Computer Vision and Pattern Recognition, 123-128.

Nicolin, B. and Gabler, R. (1987) "A Knowledge-Based System for the Analysis of Aerial Images", IEEE Trans. Geosci. Rem. Sens., GE-25 (3), 317-328.

NOAA, NOAA Polar Orbiter Data User's Guide (February 1984) U.S. Department of Commerce, National Oceanic and Atmospheric Administration.

NESDIS Polar Research Board (1984) "The Polar Regions and Climatic Change, National Research Council, National Academy Press, Washington, D. C., 59 pp.

Rabindra, P., Sengupta, S.K. and Welch, R.M. (1987) "An Interactive Hybrid Expert System for Polar Cloud and Surface Classification, Environmetrics, 3(2), 121-147.

Ripple, W.J. and Ulshoefer, V.S. (1987) "Expert Systems and Spatial Data Models for Efficient Geographic Data Handling", Photogram. Eng. Rem. Sens., 53 (10), 1431-1433.

Ritter, N.D., Logan, T.L. and Bryant, N.A. (September 1988) "Integration of Neural Network Technologies with Geographic Information Systems", GIS Symposium: Integrating Technology and Geoscience Applications, Denver, Colorado, 102-103.

Robinson, V.B. and Frank, A.U. (1985) "Expert Systems for Geographic Information Systems", Photogram. Eng. Rem. Sens., 53 (10), 1435-1441.

Robock, A., and Kaiser, D. (1985) "Satellite-Observed Reflectance of Snow and Clouds", Mon. Weather Rev., 113(11), 2023-2029.
Rossow, W. B. (1989) "Measuring Cloud Properties from Space: A Review", J. Climate, 2, 201-213.
Rossow, W. B., Mosher, F., Kinsella, E., Arking, A, Desbois, M., Harrison, E., Minnis, P., Ruprecht, E., Seze, G., Simmer, C. and Smith, E. (1985) "ISCCP Cloud Algorithm Intercomparison", J. Climate Appl. Meteorol., 24, 877-903.
Rumelhart, D.E., McClelland, J.L. and the PDP Research Group (1986) "Parallel Distributed Processing", Cambridge, MA: MIT Press, 547 pp.
Sahoo, P.K., Soltani, S., Wong, A.K.C., and Chen, Y.C. (1988) "A Survey of Thresholding Techniques, Computer Vision, Graphics, and Image Processing", 41, 233-260.
Saltzman, B., and Moritz, R. E. (1989) "A Time-Dependent Climatic Feedback System Involving Sea-Ice Extent", Ocean Temperature, and CO2, Tellus, 32, 93-118.
Schowengerdt, R.A. and Wang, H. (1989) "A General Purpose Expert System for Image Processing, Photogram. Eng. Remote Sens., 55 (9), 1277-1284.
Schwalb, A. (1984) "The TIROS-N/NOAA A-G Satellite Series", NOAA Tech. Mem., NESS95.
Shine, K. P., and Crane, R.G. (1984) "The Sensitivity of a One-Dimensional Thermodynamic Sea Ice Model to Changes of Cloudiness", J. Geophys. Res., 89(C6), 10,615-10, 622.
Skidmore, A.K. (1989) "An Expert System Classifies Eucalypt Forest Types Using Thematic Mapper Data and a Digital Terrain Model", Photogram. Eng. Remote Sens., 55 (10), 1449-1464.
Smith, T.R. (1984) "Artificial Intelligence and its Applicability to Geographical Problem Solving", Professional Geographer, 36 (2), 147-158.
Smith, T., Peuquet, D., Menon, S., and Agarwal, P. (1987) "KBGIS II: A Knowledge-Based Geographical Information System", Int. J. Geog. Sys., 1 (2), 149-172.
Stansfield, S. (1986) "ANGY: A Rule-Based Expert System for Automatic Segmentation of Coronary Vessels From Digital Subtracted Angiograms", IEEE Trans. Pattern Anal. Mach. Intell., PAMI-8 (2), 188-199.
Swain, P.H., Vardeman, S.B. and Tilton, J.C. (1981) "Contextual Classification of Multispectral Image Data", Pattern Recognition, 13, 429-441.
Tovinkere, V.R., Penaloza, M., Logar, A., Lee, J., Weger, R.C., Berendez, T.A. and Welch, R.M. (1992) "An Intercomparison of Artificial Intelligence Approaches for Polar Scene Identification", J. Geophys. Res.
Usery, E.L. and Altheide, P. (1988) "Knowledge-Based GIS Techniques Applied to Geological Engineering", Photogram. Eng. Rem. Sens., 54 (11), 1623-1628.
Wang, S., Elliot, D.B., Campbell, J.B., Erich, R.W. and Haralick, R.M. (1983) "Spatial Reasoning in Remotely Sensed Data", IEEE Trans. Geosci. Remote Sens., GE-21, 94-101.
Wang, F. and Newkirk, R. (1988) "A Knowledge-Based System for Highway Network Extraction", IEEE Trans. Geosci. Remote Sens., 26 (5), 525-531.

Welch, R. M., Sengupta, S.K., and Chen, D.W. (1988) "Cloud Field Classification Based Upon High Spatial Resolution Textural Features, 1, Gray Level Co-Occurrence Matrix Approach", J. Geophys. Res., 93(D10), 12, 663-12, 681.

Welch, R.M., Kuo, K.S. and Sengupta, S.K. (1989a) "Cloud Field Classification Based Upon High Spatial Resolution Textural", Part 3: Discrimination of Cloud and Surface Features in Polar Regions from Digitized Imagery.

Welch, R.M., Navar, M.S. and Sengupta, S.K. (1989b) "Cloud Field Classification Based Upon High Spatial Resolution Textural Features", Part 4: The Effect of Spatial Resolution, J. Geophys. Res.

Welch, R.M., Sengupta, S.K., Goroch, A.K., Rabindra, P., Rangaraj, N. and Navar, M.S. (1992) "Polar Cloud and Surface Classification Using VHRR Imagery: an Intercomparison of Methods", J. Appl. Meteorol. 31(5), 405-420.

Weszka, J.S., Dyer, C.R. and Rosenfield, A. (1976) "A Comparative Study of Texture Measures for Terrain Classification", IEEE Trans. Syst. Man. Cybern., SMC-6, 269-285.

World Meteorological Organization (WMO) (1987) "Report of the ISCCP Workshop on Cloud Algorithms in the Polar Regions", Tokyo, Japan, 19-21, August 1986, World Clim. Res. Prog., Rep. WCP-131, WMO/TD 170, 19-21 August 1986, Geneva.

World Meteorological Organization (WMO) (May 13 to June 3, 1988) "Report of the third session of the working group on sea ice and climate", Rep. WCRP-18, WMO/TD 272, Oslo, Norway.

Wu, R., Weinman, J.A. and Chin, R.T. (1985) "Determination of Rainfall Rates from GOES Satellite Images by a Pattern Recognition Technique", J. Atmos. Ocean. Tech., 2, 314-330.

Jeffrey R. Key, Cooperative Institute for Research in Environmental Sciences, University of Colorado at Boulder, Boulder, CO 80309, U.S.A.

Appendix I

NEURAL NETWORK RESOURCES

The material in this appendix has been culled extensively from the FAQ (Frequently Asked Questions) distributed on the comp.ai.neural-nets category of Netnews on Internet--the global computing network. The complete FAQ file is compiled by Lutz Prechelt (email: prechelt@ira.uka.de) with contibution by numerous others and provides a mine of information on resources and answers to questions commonly asked by newcomers to the field. The FAQ represent a consensus of information from many active researchers in the field of neural nets, and as such could be viewed as the collective wisdom of knowledgable people on issues of importance to newcomers. Readers who have access would be well advised to obtain a copy.

Most of the material provided in the FAQ assumes knowledge how to access Netnews and use FTP (File Transfer Protocol) on internet. The latest version of the FAQ is posted on Netnews once a month and may be obtained by anonymous FTP from:

host: rtfm.mit.edu
directory: /pub/usenet/news.answers

It may also be obtained via email by sending an email message with the word 'help' in the body of the message to mail-server@rtfm.mit.edu. What follows here is a selection of items from the FAQ of use to those who do not necessarily have access to internet.

Beginner introductory books - Neural Network Theory

Aleksander, I. and Morton, H. (1990) "An Introduction to Neural Computing," Chapman and Hall.

Beale, R. and Jackson, T. (1990) "Neural Computing, an Introduction," Adam Hilger, IOP Publishing Ltd : Bristol.

Dayhoff, J. E. (1990) "Neural Network Architectures: An Introduction," Van Nostrand Reinhold: New York.

McClelland, J. L. and Rumelhart, D. E. (1988) "Explorations in Parallel Distributed Processing: Computational Models of Cognition and Perception." The MIT Press.

B. C. Hewitson and R. G. Crane (eds.), Neural Nets: Applications in Geography, 181–185.

McCord Nelson, M. and Illingworth, W.T. (1990) "A Practical Guide to Neural Nets," Addison-Wesley Publishing Company, Inc.

Wasserman, P. D. (1989) "Neural Computing: Theory & Practice," Van Nostrand Reinhold: New

Introductory journal articles

Hinton, G. E. (1989) "Connectionist learning procedures," Artificial Intelligence, Vol. 40, pp. 185--234.

Knight, K. (1990) "Connectionist, Ideas and Algorithms," Communications of the ACM. Nov. 1990. Vol.33 nr.11, pp 59-74.

Dedicated Neural Network journals

Title: Neural Networks
Publish: Pergamon Press
Address: Pergamon Journals Inc., Fairview Park, Elmsford, New York 10523, USA and Pergamon Journals Ltd. Headington Hill Hall, Oxford OX3, 0BW, England

Title: Neural Computation
Publish: MIT Press
Address: MIT Press Journals, 55 Hayward Street Cambridge, MA 02142-9949, USA, Phone: (617) 253-2889

Title: IEEE Transaction on Neural Networks
Publish: Institute of Electrical and Electronics Engineers (IEEE)
Address: IEEE Service Cemter, 445 Hoes Lane, P.O. Box 1331, Piscataway, NJ, 08855-1331 USA. Tel: (201) 981-0060

Title: International Journal of Neural Systems
Publish: World Scientific Publishing
Address: USA: World Scientific Publishing Co., 687 Hartwell Street, Teaneck, NJ 07666. Tel: (201) 837-8858; Eurpoe: World Scientific Publishing Co. Pte. Ltd., 73 Lynton Mead, Totteridge, London N20-8DH, England. Tel: (01) 4462461; Other: World Scientific Publishing Co. Pte. Ltd., Farrer Road, P.O. Box 128, Singapore 9128. Tel: 2786188

Title: Neural Network News
Publish: AIWeek Inc.
Address: Neural Network News, 2555 Cumberland Parkway, Suite 299, Atlanta, GA 30339 USA. Tel: (404) 434-2187

Title: Network: Computation in Neural Systems
Publish: IOP Publishing Ltd
Address: Europe: IOP Publishing Ltd, Techno House, Redcliffe Way, Bristol BS1 6NX, UK; IN USA: American Institute of Physics, Subscriber Services 500 Sunnyside Blvd., Woodbury, NY 11797-2999

Title: Connection Science: Journal of Neural Computing, Artificial Intelligence and Cognitive Research
Publish: Carfax Publishing
Address: Europe: Carfax Publishing Company, P. O. Box 25, Abingdon, Oxfordshire OX14 3UE, UK. USA: Carafax Publishing Company, 85 Ash Street, Hopkinton, MA 01748

Neural Network Associations

International Neural Network Society (INNS).
Address: INNS Membership, P.O. Box 491166, Ft. Washington, MD 20749.

International Student Society for Neural Networks (ISSNNets).
Address: ISSNNet, Inc., P.O. Box 15661, Boston, MA 02215 USA

Women In Neural Network Research and technology (WINNERS).
Address: WINNERS, c/o Judith Dayhoff, 11141 Georgia Ave., Suite 206, Wheaton, MD

Other sources of information about NNs

Neuron Digest -- Internet Mailing List.
To subscribe, send email to neuron-request@cattell.psych.upenn.edu
Neural ftp archive site: funic.funet.fi

Some free Neural Net software

Rochester Connectionist Simulator
anonymous FTP from cs.rochester.edu (192.5.53.209)
directory : pub/simulator

UCLA-SFINX
ftp 131.179.16.6 (retina.cs.ucla.edu)
Name: sfinxftp
Password: joshua
directory: pub/

NeurDS
ftp gatekeeper.dec.com
directory: pub/DEC
file: NeurDS031.tar.Z

PlaNet5.7 (also known as SunNet)
ftp boulder.colorado.edu
directory: pub/generic-sources
file: PlaNet5.7.tar.Z

Mactivation
ftp bruno.cs.colorado.edu
directory: /pub/cs/misc
file: Mactivation-3.3.sea.hqx

SNNS
ftp: ftp.informatik.uni-stuttgart.de
directory: /pub/SNNS
file: SNNSv3.0.tar.Z
manual: SNNSv2.1.Manual.ps.Z

Aspirin/MIGRAINES
ftp: ftp.cognet.ucla.edu
directory: /alexis
file: am6.tar.Z

NeuralShell
ftp: quanta.eng.ohio-state.edu
directory: pub/NeuralShell
file: NeuralShell.tar

Xerion
ftp: ftp.cs.toronto.edu
directory: /pub/xerion.
file: xerion-3.1.ps.Z

LVQ_PAK, SOM_PAK
ftp: cochlea.hut.fi
directory /pub/lvq_pak and /pub/som_pak
files: lvq_pak-2.1.tar.Z som_pak-1.1.tar.Z

Nevada Backpropagation (NevProp)
ftp: unssun.scs.unr.edu
directory: pub/goodman/nevpropdir

Appendix II

FORTRAN 77 LISTING FOR KOHONEN SELF ORGANIZING MAP

```
C*koh.f *spatial classification by 2D Self Organising Map (2D-SOM)
C* Stan Openshaw, June 1993, School of Geography, Leeds University
C* Leeds LS2 9JT, UK email: stan@geog.leeds.ac.uk

C++++++++++++++++++++++++++++++++++++++++
C* NVAR= Number of Variables
C* NCASE= Number of cases to be classified
C* MAXGDE= Maximum number of rows in net
C* MAXGDN= Maximum number of cols in net
C* MAXIT= Maximum number of training cases
C* LAT= 0 for block party, =1 for Gaussian inhibition function

C* Program currently expects 84 variables and 8630 cases to be read

C* CPU times are linear in NVAR, in MAXIT, and in size of map
C* This run took three hours on Sunsparc10 model 30 workstation

PARAMETER (NVAR=84,NCASE=8530,MAXIT=10000000)
PARAMETER (MAXGDE=5,MAXGDN=5,LAT=0)

C++++++++++++++++++++++++++++++++++++

DIMENSION G(NVAR,MAXGDE,MAXGDN)
DIMENSION X(NVAR,NCASE),XM(NVAR),WEI(NCASE),
     VALUES(NVAR)
INTEGER LIST(NCASE),ICL(NCASE),NUM(NCASE) CHARACTER*10 FILE7

C++++++++++++++++++++++++++++++++++++
C* set output file
FILE7 = 'som.res'

C* Report run parameters
WRITE (6,78123) NCASE,NVAR,MAXGDE,MAXGDN,MAXIT

78123 FORMAT ('*Number of Cases to be read',
+ I6/'*Number of Variables to be read',
+ I6/'*Two-Dimensional SOM has',I4,' rows and',I5,
+ ' columns'/'*Number of Training cases ',I10)

IF (LAT.EQ.0) WRITE (6,78231)
```


B. C. Hewitson and R. G. Crane (eds.), Neural Nets: Applications in Geography, 187–194.

```
78231 FORMAT ('*Simple Block Party used')

IF (LAT.NE.0) WRITE (6,78232)

78232 FORMAT ('*Gaussian lateral inhibition function used')

C* set some constants to do with training to some arbitrary values
ALPHA0 = 0.4
ALPHA = ALPHA0
DSTEP0 = MAX0(MAXGDE,MAXGDN)/2.0
DSTEP = DSTEP0
DMAXTR = MAXIT

C* initialize means
DO 21 J = 1,NVAR
21 XM(J) = 0.0
IDUM = 1
XX = RAND(IDUM)
IDUM = 0

C* read data and store
IRC = 1
DO 1200 I = 1,NCASE
C* read a record putting data values in VALUES(*) and weighting
C* value in WEIGHT (must be positive but can be real number)
CALL INPUT1(VALUES,WEIGHT,NVAR,IRC)
IF (IRC.NE.0) GO TO 1300
DO 1250 J = 1,NVAR
XM(J) = XM(J) + VALUES(J)
1250 X(J,I) = VALUES(J)
WEI(I) = ABS(WEIGHT)
1200 CONTINUE
WRITE (6,7811)

7811 FORMAT ('*All Data Read ok')

C* Calc variable means
DO 22 J = 1,NVAR
22 XM(J) = XM(J)/NCASE

C* init map weights
DO 110 K = 1,NVAR
DO 110 I = 1,MAXGDN
DO 110 J = 1,MAXGDE
110 G(K,J,I) = ((RAND(IDUM)-0.5)*XM(K))/3.0 + XM(K)

C* Other inits and form cumulative sampling sums
ICL(1) = 0
DO 10 I = 2,NCASE
ICL(I) = 0
```

```
10 WEI(I) = WEI(I) + WEI(I-1)

C* START ITERATIONS===================== MOVERS = 0
KIK = 0
NUP = 0

SUMSAM = WEI(NCASE)
WRITE (6,71) SUMSAM

71 FORMAT ('*Sum of Weights=',G15.9)

DCON = 2.0/DSTEP/DSTEP

AVER = 0.0

DO 1000 ITER1 = 1,MAXIT


C* select training case at random but proportional to weighting TARGET
= RAND(IDUM)*SUMSAM

C* perform binary search to find selected case
MAX = NCASE
MIN = 1
K = (MAX+MIN)/2
4000 CONTINUE
IF (TARGET.EQ.WEI(K)) THEN
KK = K
GO TO 5000

END IF

IF (WEI(K).GT.TARGET) THEN
MAX = K - 1

ELSE
MIN = K + 1
END IF
IF (MAX.GE.MIN) THEN
K = (MIN+MAX)/2
GO TO 4000

END IF

KK = MIN
5000 CONTINUE

C* Record=KK to be used, so copy its data
C* It could be a direct access read
DO 551 J = 1,NVAR
551 VALUES(J) = X(J,KK)
```

```
C* call variable randomiser to allow each variable to
C* possess a different amount of measurement/precision
C* reliability
CALL VARRAN(VALUES,NVAR)

C* Find best matching neuron
BEST = 1E30
DO 150 J = 1,MAXGDN
DO 150 K = 1,MAXGDE

C* Use sum of absolute differences as measure of dissimilarity
SUM = 0.0
DO 160 L = 1,NVAR
160 SUM = SUM + ABS(G(L,K,J)-VALUES(L))
IF (SUM.GE.BEST) GO TO 150
BEST = SUM
INDJ = J
INDK = K
150 CONTINUE

AVER = AVER + BEST
C* Set classification
NEW = (INDJ-1)*MAXGDE + INDK
IF (NEW.NE.ICL(KK)) THEN
MOVERS = MOVERS + 1
ICL(KK) = NEW
END IF
C* set distance threshold for block neighborhood
DSQ = DSTEP*DSTEP

DCON = 2.0/DSQ
C* Now update best matched neuron AND its neighbors
DO 200 J = 1,MAXGDN
DO 200 K = 1,MAXGDE

C* Calc distance from winning neuron
DIS = (J-INDJ)**2 + (K-INDK)**2

C* compute weights due to lateral inhibition
IF (LAT.EQ.0) THEN
IF (DIS.GT.DSQ) GO TO 200
SYN = ALPHA

ELSE
GAUS = DIS/DCON
IF (GAUS.GT.20.0) THEN
GAUS = 0.0

ELSE
GAUS = EXP(-GAUS)
```

```
END IF

SYN = ALPHA*GAUS
END IF

C* update weights
IF (SYN.NE.0.0) THEN
NUP = NUP + 1
DO 320 L = 1,NVAR
320 G(L,K,J) = G(L,K,J) + SYN* (VALUES(L)-G(L,K,J))
END IF


200 CONTINUE

C* Print something every now and again
KIK = KIK + 1
IF (KIK.EQ.10000) THEN
AVER = AVER/10000.0
WRITE (6,32) ITER1,MOVERS,DSTEP,ALPHA,NUP,AVER
MOVERS = 0
KIK = 0
AVER = 0.0
NUP = 0
END IF

32 FORMAT ('*Iteration No',I8,' Moves=',I8,' DSTEP=',G15.9,
+ ' ALPHA=',G15.9,' UPDATES=',I9,' Aver Error=',G15.9)

C* Change training parameters
ALPHA = ALPHA0* (1.0-ITER1/DMAXTR)
DSTEP = DSTEP0* (1.0-ITER1/DMAXTR)

1000 CONTINUE

C========END of TRAINING===============

OPEN (UNIT=7,FILE=FILE7,FORM='FORMATTED',
    STATUS='UNKNOWN')
C*Output Neuron Weights
DO 333 J = 1,MAXGDN
DO 333 K = 1,MAXGDE
333 WRITE (7,334) K,J, (G(L,K,J),L=1,NVAR)

334 FORMAT (2I4/ (10F12.4))

C* Classify data
DO 500 I = 1,NCASE

C* Find nearest neuron for I
BEST = 1E46
```

```
DO 502 J = 1,MAXGDN
DO 502 K = 1,MAXGDE

C* Calc dissimilarity
SUM = 0.0
DO 503 L = 1,NVAR
503 SUM = SUM + ABS(G(L,K,J)-X(L,I))
IF (SUM.GE.BEST) GO TO 502
BEST = SUM
INDJ = J
INDK = K
502 CONTINUE

C* Save result

LIST(I) = (INDJ-1)*MAXGDE + INDK
500 CONTINUE

C* Normalize classification
NC = 0
DO 521 I = 1,NCASE
IND = LIST(I)
IF (IND.LT.0) GO TO 521
NC = NC + 1
NUM(NC) = 0
DO 522 J = 1,NCASE
IF (LIST(J).EQ.IND) THEN
ICL(J) = NC
LIST(J) = -1
NUM(NC) = NUM(NC) + 1
END IF

522 CONTINUE
521 CONTINUE

WRITE (6,7823) NC, (J,NUM(J),J=1,NC)

7823 FORMAT ('*Number of clusters=',I6/
+ (5X,'Cluster No.',I4,' has ',I6,'members'))

C* write out classification
C* I-th case assigned to cluster ICL(I)
WRITE (7,19008) ICL

19008 FORMAT (13I6)

CLOSE (UNIT=7,STATUS='KEEP')

STOP

C* errors
```

```
1300 WRITE (6,1301)

1301 FORMAT ('******ERROR: Unexpected EOF during data input')

STOP 1

END
C===Example Input subroutine===================
SUBROUTINE INPUT1(VALUES,WEIGHT,NVAR,IRC) DIMENSION VALUES(NVAR)
CHARACTER*10 FILE1
C* open file first time called
IF (IRC.EQ.1) THEN
FILE1 = 'sec.dat'
OPEN (UNIT=1,FILE=FILE1,FORM='FORMATTED',
STATUS='OLD')
END IF

IRC = 0
C* read data record
READ (1,1201,END=999) WEIGHT,VALUES

1201 FORMAT (T41,F8.0,84F10.2)

RETURN
C* close file on eof
999 CLOSE (UNIT=1,STATUS='KEEP')
IRC = 1
WRITE (6,21)

21 FORMAT ('NCASE too large?')

RETURN
END

C===Example Variable noise subroutine==============
C* User supplied routine to add noise to variables to reflect
C* measurement error/precision variations
SUBROUTINE VARRAN(VALUES,NVAR)
DIMENSION VALUES(NVAR)
C* This allows you to allow the SOM to cope with spatial data
C* that is possessed of unequal levels of measurement error
C* (ie a mix of 10% and 100% census data). Differing
C* levels of precision due to differences in area size
C* are best handled by the weight variable and need not be
C* included here.

C* An example, would be to compute sampling errors due to
C* simple random sampling and generate normally distributed
C* random values with a mean equivalent to the observed data
C* and standard deviation according to sampling error estimate.
```

```
C* Different variables may (quite naturally) have different
C* levels of noise and this can be catered for. You should
C* aim to be realistic rather than massively accurate.  Something
C* is better than nothing at all.
RETURN

END
```

The GeoJournal Library

1. B. Currey and G. Hugo (eds.): *Famine as Geographical Phenomenon.* 1984 ISBN 90-277-1762-1
2. S. H. U. Bowie, F.R.S. and I. Thornton (eds.): *Environmental Geochemistry and Health.* Report of the Royal Society's British National Committee for Problems of the Environment. 1985 ISBN 90-277-1879-2
3. L. A. Kosiński and K. M. Elahi (eds.): *Population Redistribution and Development in South Asia.* 1985 ISBN 90-277-1938-1
4. Y. Gradus (ed.): *Desert Development.* Man and Technology in Sparselands. 1985 ISBN 90-277-2043-6
5. F. J. Calzonetti and B. D. Solomon (eds.): *Geographical Dimensions of Energy.* 1985 ISBN 90-277-2061-4
6. J. Lundqvist, U. Lohm and M. Falkenmark (eds.): *Strategies for River Basin Management.* Environmental Integration of Land and Water in River Basin. 1985 ISBN 90-277-2111-4
7. A. Rogers and F. J. Willekens (eds.): *Migration and Settlement.* A Multiregional Comparative Study. 1986 ISBN 90-277-2119-X
8. R. Laulajainen: *Spatial Strategies in Retailing.* 1987 ISBN 90-277-2595-0
9. T. H. Lee, H. R. Linden, D. A. Dreyfus and T. Vasko (eds.): *The Methane Age.* 1988 ISBN 90-277-2745-7
10. H. J. Walker (ed.): *Artificial Structures and Shorelines.* 1988 ISBN 90-277-2746-5
11. A. Kellerman: *Time, Space, and Society.* Geographical Societal Perspectives. 1989 ISBN 0-7923-0123-4
12. P. Fabbri (ed.): *Recreational Uses of Coastal Areas.* A Research Project of the Commission on the Coastal Environment, International Geographical Union. 1990 ISBN 0-7923-0279-6
13. L. M. Brush, M. G. Wolman and Huang Bing-Wei (eds.): *Taming the Yellow River: Silt and Floods.* Proceedings of a Bilateral Seminar on Problems in the Lower Reaches of the Yellow River, China. 1989 ISBN 0-7923-0416-0
14. J. Stillwell and H. J. Scholten (eds.): *Contemporary Research in Population Geography.* A Comparison of the United Kingdom and the Netherlands. 1990 ISBN 0-7923-0431-4
15. M. S. Kenzer (ed.): *Applied Geography.* Issues, Questions, and Concerns. 1989 ISBN 0-7923-0438-1
16. D. Nir: *Region as a Socio-environmental System.* An Introduction to a Systemic Regional Geography. 1990 ISBN 0-7923-0516-7
17. H. J. Scholten and J. C. H. Stillwell (eds.): *Geographical Information Systems for Urban and Regional Planning.* 1990 ISBN 0-7923-0793-3
18. F. M. Brouwer, A. J. Thomas and M. J. Chadwick (eds.): *Land Use Changes in Europe.* Processes of Change, Environmental Transformations and Future Patterns. 1991 ISBN 0-7923-1099-3

The GeoJournal Library

19. C. J. Campbell: *The Golden Century of Oil 1950–2050.* The Depletion of a Resource. 1991 ISBN 0-7923-1442-5
20. F. M. Dieleman and S. Musterd (eds.): *The Randstad: A Research and Policy Laboratory.* 1992 ISBN 0-7923-1649-5
21. V. I. Ilyichev and V. V. Anikiev (eds.): *Oceanic and Anthropogenic Controls of Life in the Pacific Ocean.* 1992 ISBN 0-7923-1854-4
22. A. K. Dutt and F. J. Costa (eds.): *Perspectives on Planning and Urban Development in Belgium.* 1992 ISBN 0-7923-1885-4
23. J. Portugali: *Implicate Relations.* Society and Space in the Israeli-Palestinian Conflict. 1993 ISBN 0-7923-1886-2
24. M. J. C. de Lepper, H. J. Scholten and R. M. Stern (eds.): *The Added Value of Geographical Information Systems in Public and Environmental Health.* 1994 ISBN 0-7923-1887-0
25. J. P. Dorian, P. A. Minakir and V. T. Borisovich (eds.): *CIS Energy and Minerals Development.* Prospects, Problems and Opportunities for International Cooperation. 1993 ISBN 0-7923-2323-8
26. P. P. Wong (ed.): *Tourism vs Environment: The Case for Coastal Areas.* 1993 ISBN 0-7923-2404-8
27. G. B. Benko and U. Strohmayer (eds.): *Geography, History and Social Sciences.* 1994 ISBN 0-7923-2543-5
28. A. Faludi and A. der Valk: *Rule and Order. Dutch Planning Doctrine in the Twentieth Century.* 1994 ISBN 0-7923-2619-9
29. B. C. Hewitson and R. G. Crane (eds.): *Neural Nets: Applications in Geography.* 1994 ISBN 0-7923-2746-2

KLUWER ACADEMIC PUBLISHERS – DORDRECHT / BOSTON / LONDON

The manufacturer's authorised representative in the EU is Springer Nature Customer Service Centre GmbH, Europaplatz 3, 69115 Heidelberg, Germany. If you have any concerns regarding our products, please contact ProductSafety@springernature.com

Printed and bound by CPI Group (UK) Ltd, Croydon, CR0 4YY
23/04/2026
02095623-0004